Venus, closest planet to the Earth, is a torrid world of extremes shrouded from direct view by dense clouds. This, *Atlas of Venus*, shows all the fascinating detail discovered on the recent Magellan mission to map the planet surface.

Giving the historical background to our perception of the planet, this book clearly explains why Venus has been the goal of so many missions by both Russian and American Space programs. With the latest images from the Magellan mission, this colourful *Atlas* shows the beautiful landscape of Venus and its dynamic volcanism. Over 100 maps and illustrations in this *Atlas* show the dramatic beauty of this photogenic planet.

Complete with detailed maps of the planet and a gazetteer of all landmarks this is the essential reference account for all professional and amateur astronomers, and planetary scientists interested in our closest neighbour.

Atlas of Venus

ATLAS

OF VENUS

PETER CATTERMOLE AND PATRICK MOORE

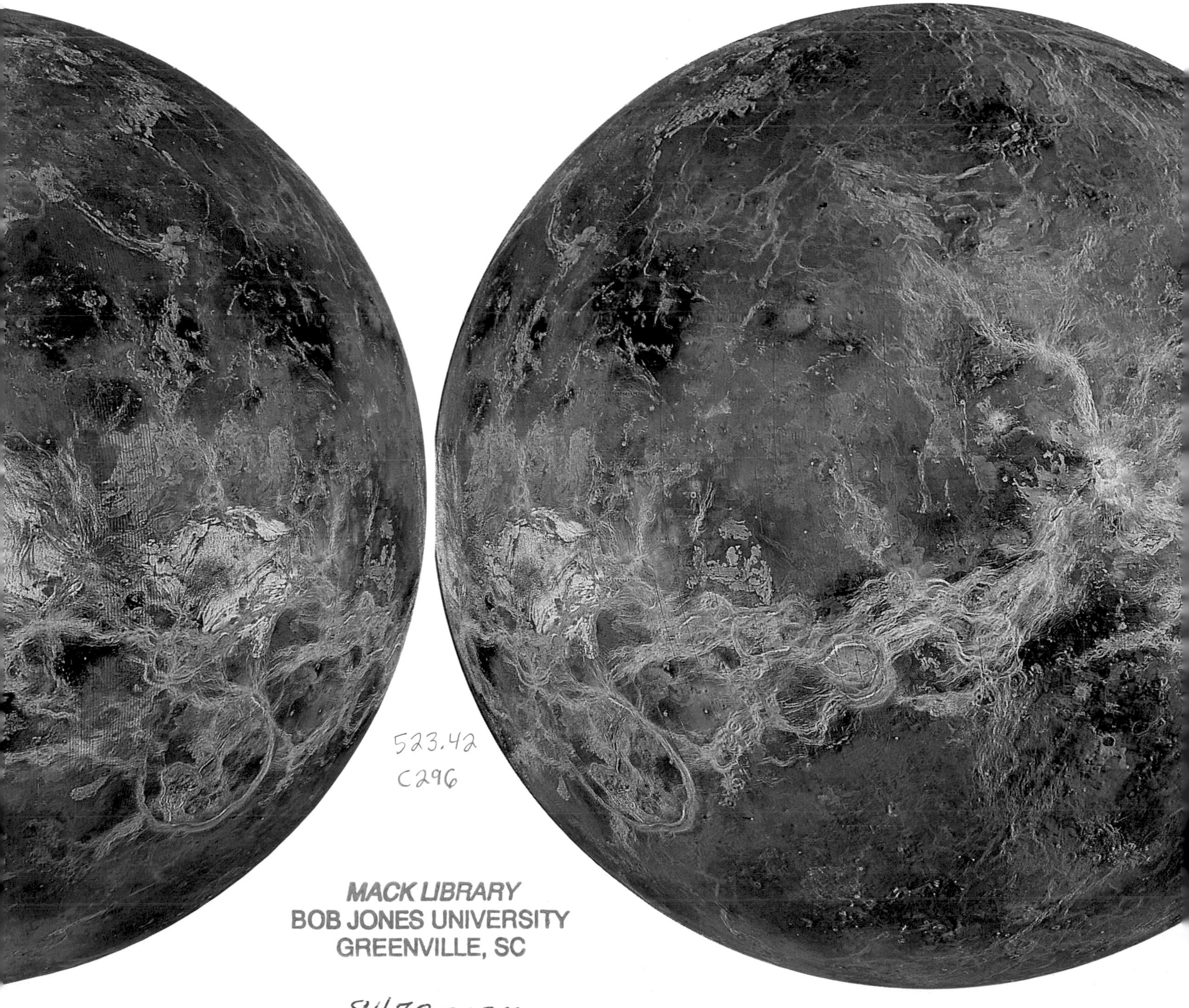

PUBLISHED BY THE PRESS SYNDICATE OF THE UNIVERSITY OF CAMBRIDGE
The Pitt Building, Trumpington Street, Cambridge CB2 1RP, United Kingdom

CAMBRIDGE UNIVERSITY PRESS
The Edinburgh Building, Cambridge CB2 2RU, United Kingdom
40 West 20th Street, New York, NY 10011–4211, USA
10 Stamford Road, Oakleigh, Melbourne 3166, Australia

First published 1997

Printed in Italy by Rotolito Lombarda

Typeset in Adobe Caslon 10/12pt and Monotype Grotesque

A catalogue record for this book is available from the British Library

Library of Congress Cataloguing in Publication data
Cattermole, Peter John
Atlas of Venus / Peter Cattermole and Patrick Moore.
p. cm.
Includes bibliographical references and index.
ISBN 0 521 49652 7 (hbk)
1. Venus (Planet) – Atlases. 2. Magellan (Spacecraft) I. Moore, Patrick. II. Title
QB621.C35 1997
523.4'2–dc20 96–23817 CIP

ISBN 0 521 49652 7 hardback

CONTENTS

PICTURE ACKNOWLEDGEMENTS

x–xv, 39, 45, 52–6, 58–64, 67–70, 72, 75, 77, 78, 81, 84–6, 88–102: All photographs courtesy of Jet Propulsion Laboratory, California Institute of Technology, National Aeronautics and Space Administration, Pasadena, USA.

4, 22, 24: Photographs by Patrick Moore.

29: Courtesy of H. E. Ball.

35: Courtesy of James W. Head III, Brown University, Rhode Island.

43: Courtesy of D. B. Campbell and B .A. Burns.

48: Courtesy of USSR. National Academy of Sciences.

66, 71: Courtesy of J. W. Head, L .S. Crumpler and J. C. Aubele, *Journal of Geophysical Research*, **97**, 13153–97 (1992). Copyright American Geophysical Union.

76: Courtesy of R. Greeley, R. E. Arvidson, C. Elachi, M. A. Geringer, J. J. Plaut, R. S. Saunders, G. Schubert, E. R. Stofan, E. J. P. Thouvenot, S. D. Wall and C. M. Weitz, *Journal of Geophysical Research*, **97**, 13319–24 (1992). Copyright American Geophysical Union.

82, 83: Courtesy of G. G. Schaber, R. G. Strom, H. J. Moore, L. A. Soderblom, R. L. Kirk, D. J. Chadwick, D. D. Dawson, L. R. Gaddis, J. M. Boyce and J. Russell. *Journal of Geophysical Research*, **97**, 13257–301 (1992). Copyright American Geophysical Union.

88: Courtesy of G. G. Shaber, *Geophysical Research Letters*, **9**, 499–512 (1982).

PREFACE

Several books dealing with the planet Venus have appeared in recent years, but most of these either are purely popular or else highly technical. The present book is an attempt to fill a gap in the literature, and to provide the general reader with a picture of Venus as we now know it to be.

The text was, of course, planned by the two authors in the closest consultation; in the event Patrick Moore wrote most of Chapters 1–4 and 12, and Peter Cattermole the remainder. We hope that the result will be acceptable.

May 1996

Peter Cattermole, Sheffield
Patrick Moore, Selsey

VENUS: **NORTHERN HEMISPHERE**

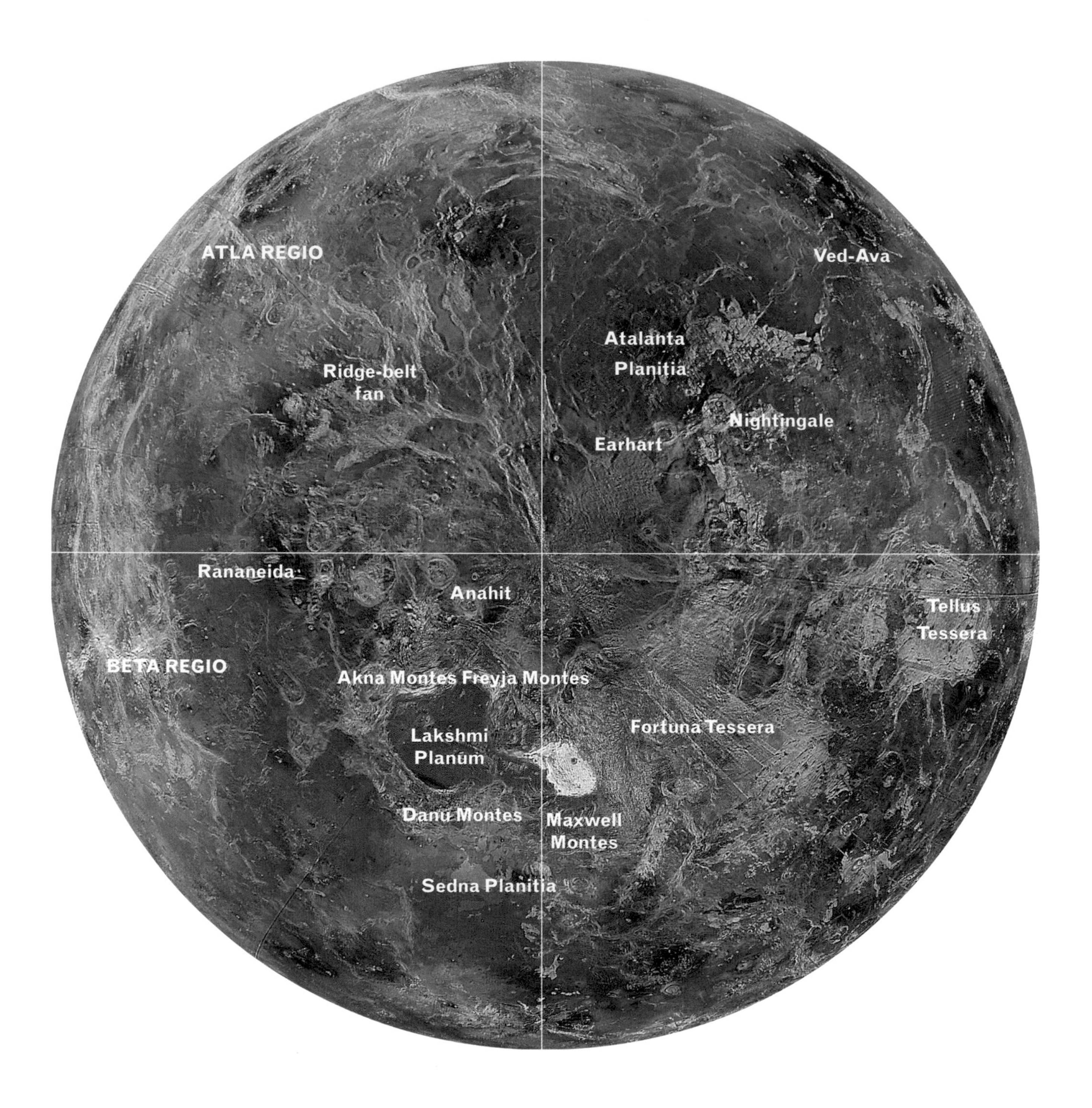

VENUS: **HEMISPHERE CENTRED AT 0° EAST LONGITUDE**

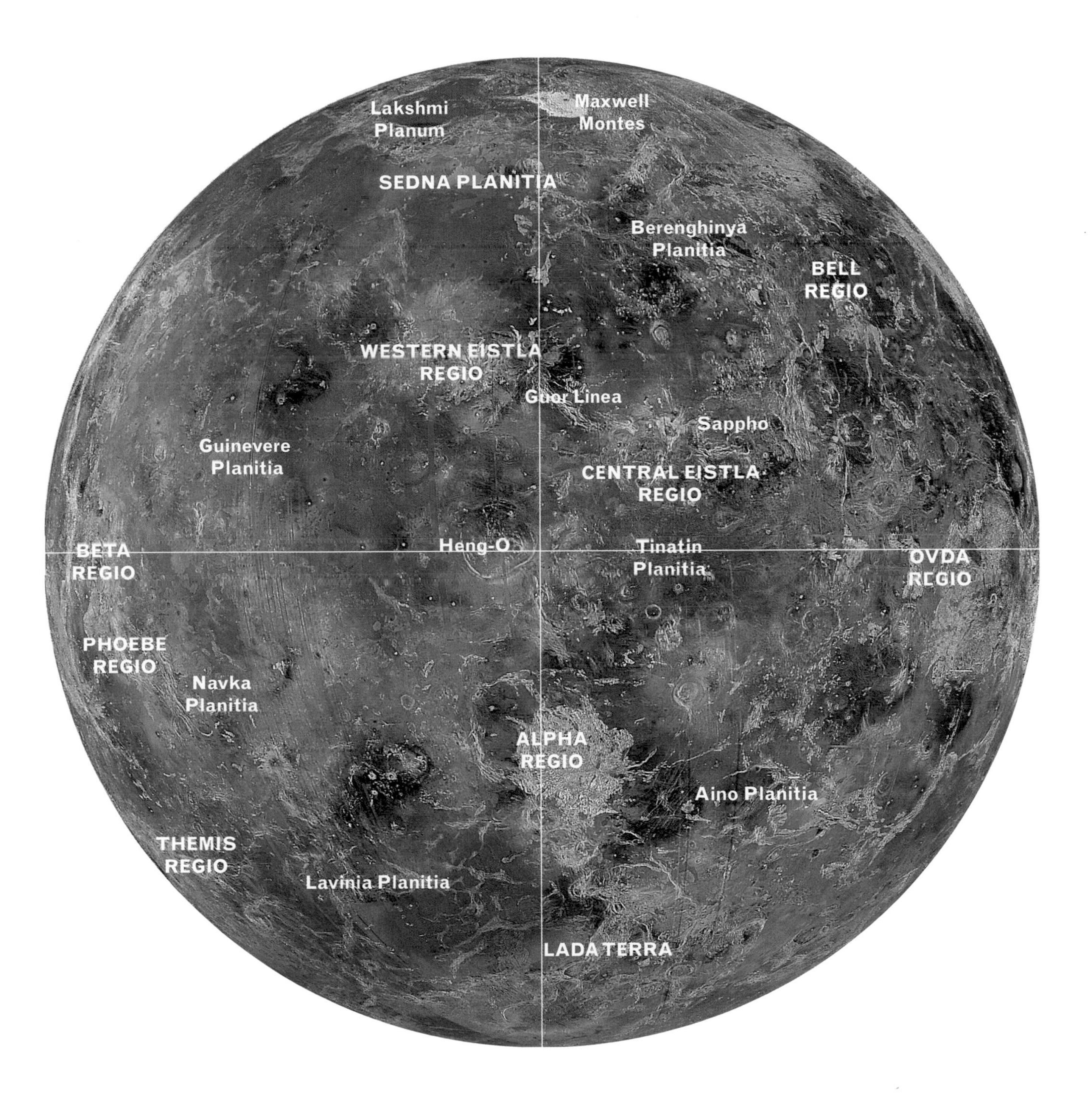

VENUS: HEMISPHERE CENTRED AT 90° EAST LONGITUDE

VENUS: **HEMISPHERE CENTRED AT 180° EAST LONGITUDE**

VENUS: **HEMISPHERE CENTRED AT 270° EAST LONGITUDE**

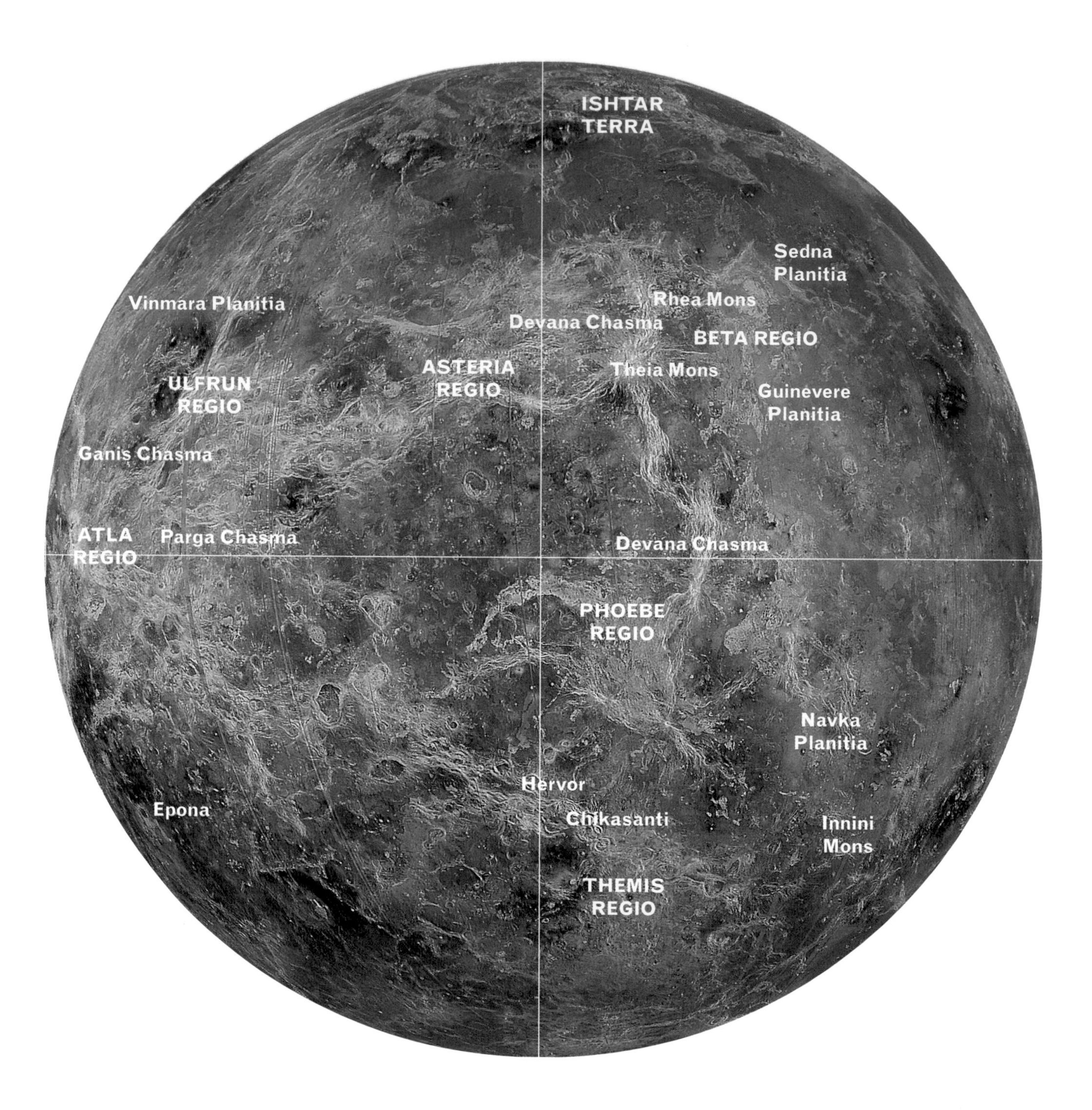

VENUS: **SOUTHERN HEMISPHERE**

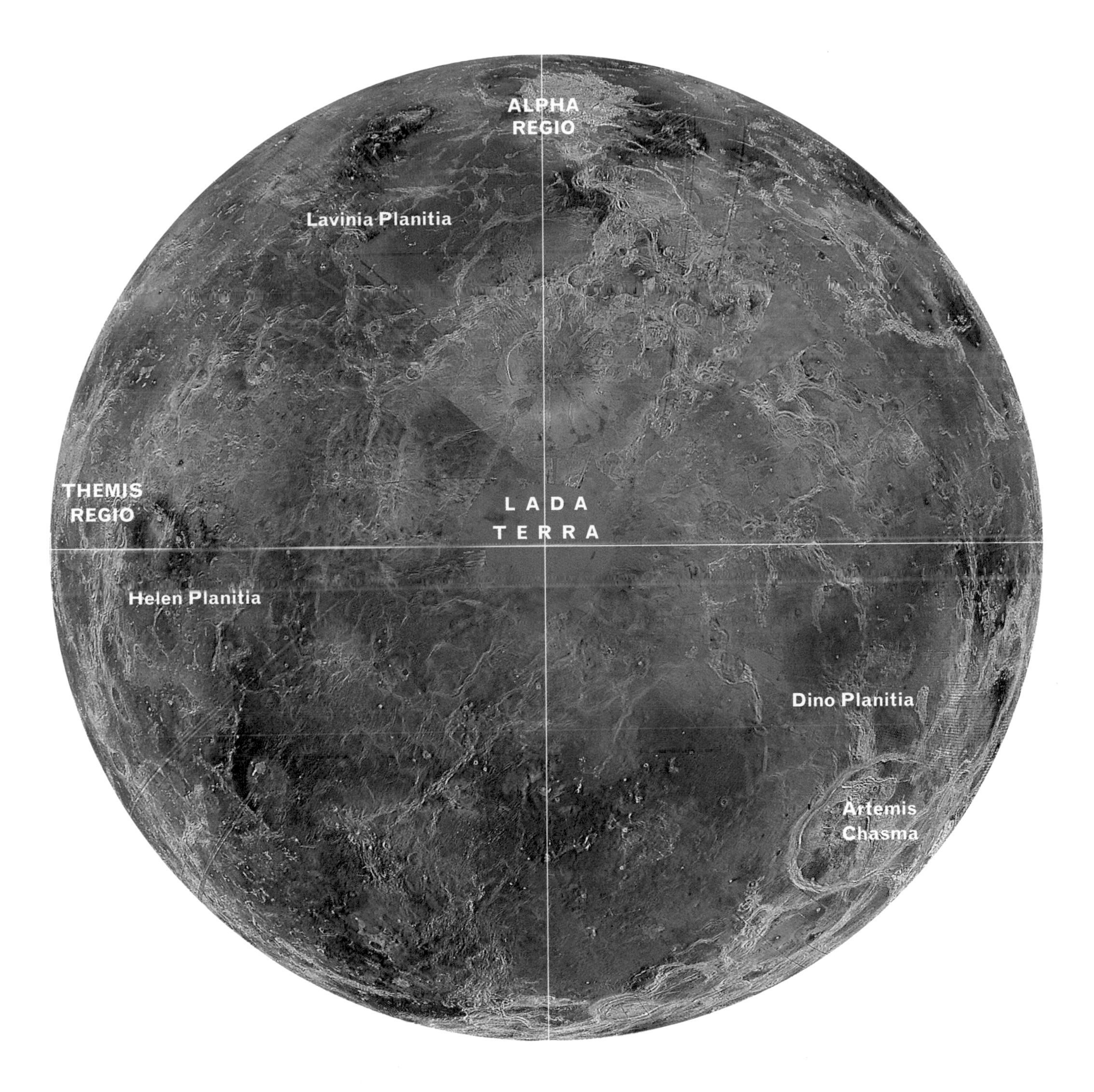

VENUS IN THE SOLAR SYSTEM

1

Venus is the planet most like the Earth. One of the four terrestrial planets, it is a dense, rocky world dominated by silicate materials, and covered by a thick mantle of carbon dioxide. So far as we know, the phases of Venus were discovered by Galileo, in 1610.

Venus, the most brilliant of the planets, must have been known since the dawn of human history. When seen shining down in the west after sunset or in the east before sunrise it may even cast shadows, and it looks almost like a small lamp in the sky, so that it is not surprising that the ancients named it in honour of the goddess of beauty and love.

What makes it particularly interesting is the fact that outwardly it is almost a twin of the Earth. It is only slightly smaller and less massive, and it has a dense atmosphere; though it is on average just over 40 million kilometres closer to the Sun than we are, there was no reason to suppose that it might be intolerably hot. Until only a few decades ago many astronomers believed that it might be largely covered in ocean, and that life might exist there.

The Solar System is divided into two well-marked parts. First, there are four relatively small, dense planets with solid surfaces: Mercury, Venus, Earth and Mars. Then comes a wide gap in which move thousands of dwarf worlds known as asteroids, which are also rocky and dense; of these only one (Ceres) is as much as 900 km in diameter, and only one (Vesta) is ever visible to the naked eye. Beyond the asteroid zone come the four giant worlds: Jupiter, Saturn, Uranus and Neptune. Of these, the former two consist largely of gas while the latter two are made up of a mixture of ices and gas. Finally there are Pluto and its companion Charon. Pluto is known to be smaller than Earth's Moon, and probably unworthy of true planetary status. The last three members of the planetary system have been discovered during the telescopic era: Uranus in 1781, Neptune in 1846 and Pluto as recently as 1930. Uranus can just be seen with the naked eye, but Neptune and Pluto are much too faint.

The Earth is actually the largest of the four inner planets. The details are shown in Table 1. The oldest observations of Venus which have come down to us are Babylonian, and are recorded on the famous Venus Tablet found by Sir Henry Layard at Konyunjik; they were deciphered in 1911 by the German scholar F. X. Kugler. When the planet appears, "rains will be in the heavens"; it is then absent for four months, and when it returns "hostility will be in the land", so that Venus was not always regarded as benevolent! Venus was called Istar, the personification of woman and the mother of the gods; temples to her were set up in Nineveh and various other places. Istar was held to be responsible for the world's fertility, and there

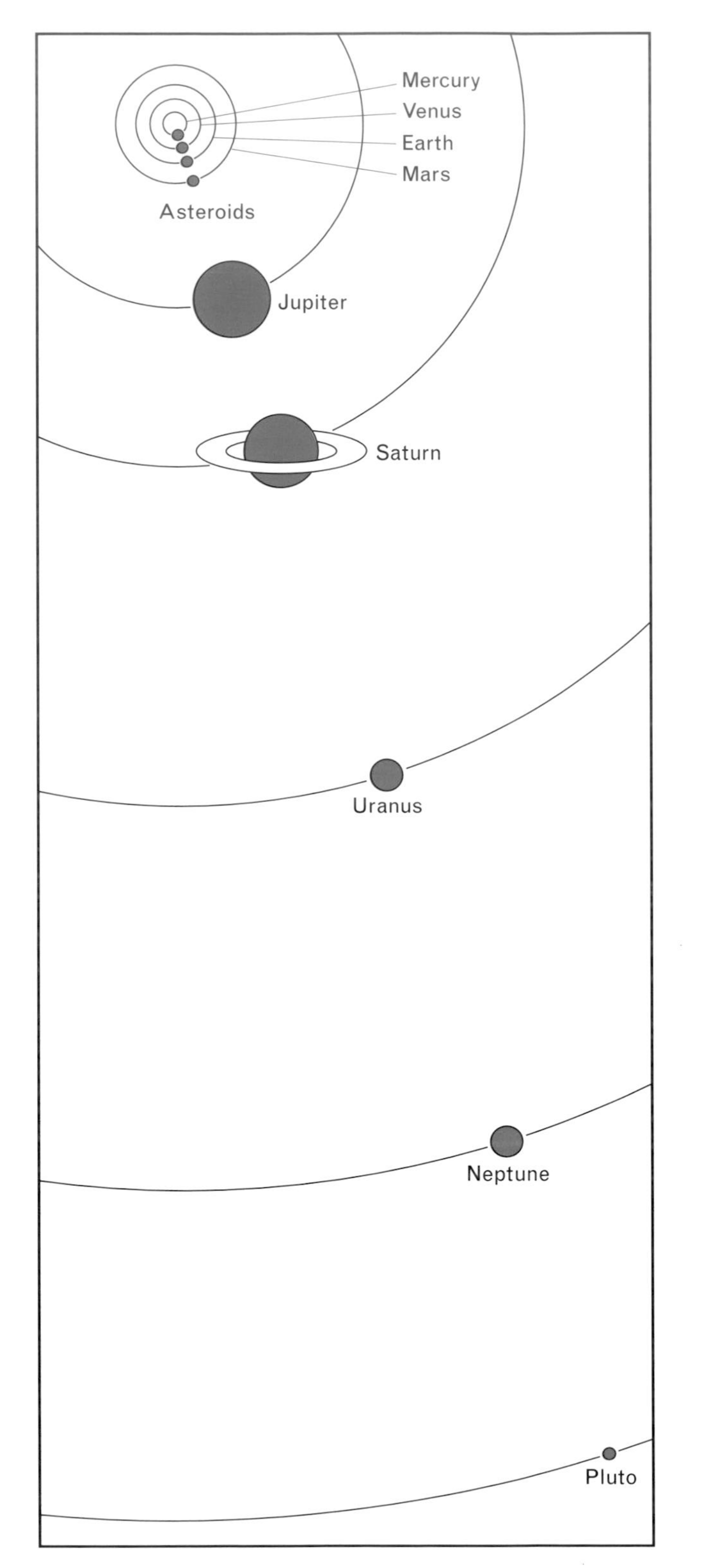

The Solar System, showing planetary orbits to scale. Note the inner and outer planet groupings.

was a legend that when she visited the Underworld to search for her dead lover Tammuz, all life on Earth began to die – to be saved only by the intervention of the gods, who revived Tammuz and so restored Istar to the world. To the Chinese, Venus was Tai-pe, or the "Beautiful White One"; in Egypt, Venus was "Bonou", the Bird, was Ouaîti as the evening star and Tioumoutiri as a morning star. At first it was no doubt believed that the evening star and the morning star were two different bodies, but at a fairly early stage it must have been realized that they were one and the same.

It was probably true to say that real astronomical science began with the Greeks, but it seems worth digressing briefly to mention some of the later civilizations. Venus was generally regarded as having female associations, except in India; the month of April was sacred to the goddess, and our name Friday is derived from the Anglo-Saxon "Frigedaeg" (Friga, or Venus, and daeg, day). But it was in the New World that Venus was regarded as particularly important. The Maya drew up an elaborate calendar, as we know from the Dresden Codex – one of the few records which have come down to us; almost all the rest were destroyed by the Spanish invaders in a deliberate attempt to wipe out the ancient culture (one of the worst of all acts of scientific vandalism). The Maya compared the synodic period of Venus (584 days) with the year (365 days), and realized that eight 365-day years are equal to five "Venus years" – 2920 days in each case, so that after this interval Venus would return to the same part of the

Table 1 The inner planets

	Mercury	Venus	Earth	Mars
Distance from Sun, 10^6 km				
Maximum	69.7	109	152	249
Mean	57.9	108.2	149	228
Minimum	45.9	107.4	147	207
Sidereal period, days	87.969	224.701	365.256	686.980
Rotation period	58.65d	243.01d	23h 56m 04s	24h 37m 23s
Axial inclination, degrees	2	178	23.44	24
Orbital eccentricity	0.206	0.007	0.017	0.093
Orbital inclination, degrees	7.0	3.4	—	1.8
Mean orbital velocity km s^{-1}	47.87	35.02	29.79	24.1
Synodic period, days	115.88	583.9	—	779.9
Diameter, km	4878	12,104	12,756 (equatorial) 12,714 (polar)	6794 (equatorial) 6750 (polar)
Reciprocal mass, Sun = 1	6,000,000	408,520	328,946	3,098,700
Mass, Earth = 1	0.055	0.815	1	0.107
Volume, Earth = 1	0.056	0.88	1	0.150
Escape velocity, km s^{-1}	4.25	10.36	11.18	5.03
Density, water = 1	5.44	5.25	5.52	3.94
Surface gravity, Earth = 1	0.38	0.90	1	0.38
Oblateness	Negligible	Negligible	0.003	0.006
Albedo	0.06	0.76	0.38	0.116
Mean surface temperature, °C	+360 (day) −170 (night)	−33 (cloud-tops) +480 (surface)	+22	−23
Maximum magnitude	−1.9	−4.4	—	−2.8
Mean diameter of Sun, seen from planet	1°22′40″	44′15″	32′01″	21″
Apparent diameter seen from Earth, sec.				
Maximum	12.9	65.2	—	25.7
Minimum	4.5	9.5	—	3.5
Atmospheric pressure at surface, bar	$\pm 10^{-10}$	90	1	0.007
Main atmospheric constituents	Helium, hydrogen	Carbon dioxide (96%), nitrogen	Nitrogen, oxygen	Carbon dioxide (95%), nitrogen

sky at the same time of the year. Then, too, there is the Caracol circular tower at the ancient capital of Chichén Itza, still to be seen as one of the world's great tourist attractions. Whether or not it functioned as any kind of astronomical observatory is a matter for debate, but it may have done, and probably it was associated with Venus – and there is little doubt that Venus was linked with the Toltec Quetzalcoatl, who was their equivalent of the Mayan Kukulcan. Note, incidentally, that some of the old Mexican peoples were afraid of Venus, and when it appeared the custom was to close doors and windows as a protection against the planet's harmful rays.

Now let us go back to the Greeks, who named the planet after Aphrodite, the goddess of beauty and love. Homer, in the Iliad, called Venus "Hesperos, which is the most beautiful star set in the sky", and the great philosophers studied its movement carefully, drawing up tables which under the circumstances were remarkably good. Note that what is often called the "Greek miracle" was not rapid. Thales, first of the philosophers, was born around 624 BC; Ptolemy, the last, died around 180 AD – an interval of over 800 years, so that in time Ptolemy was as far removed from Thales as we are from the Crusades. Still, a tremendous amount of progress was made.

The main drawback was the belief that the Earth must lie at the centre of the universe, with all the other bodies revolving around it once every 24 hours. A few far-sighted Greeks, notably Aristarchus of Samos around 280 BC, were bold enough to claim that the Sun, not the Earth, was the true centre, but there was no proof, and Aristarchus found few followers. Quite apart from any other considerations, relegating the Earth to the status of a mere planet was taken to be heretical.

The Mayan complex of Chichén Itza. This probably was related in some way to the planet Venus.

This scheme of things reached its highest perfection in the time of Ptolemy, who lived from about 120 to 180AD; we cannot be precise about these dates, and about his personality we know absolutely nothing, though it is quite definite that he spent all his life in Alexandria. Ptolemy – or, to give him his correct name, Claudius Ptolemaeus – produced a great book which has come down to us by way of its Arab translation, and is always called the Almagest. It contained a star catalogue, based upon the work of the earlier philospher Hipparchus, and is really a compendium of all the scientific knowledge of the time. Ptolemy was unquestionably a brilliant observer as well as an expert mathematician (periodic attempts to discredit him, and to claim that he was a mere copyist, have been signally unsuccessful), and we owe him a great debt of gratitude; without the Almagest we would know far less about ancient science than we actually do. Ptolemy even drew a map of the civilized world which was based on proper observation rather than guesswork.

His theory of the universe is always known as the Ptolemaic, though Ptolemy himself did not actually invent it. The Earth lies at the centre of the universe, and the various celestial bodies move round it in perfectly circular orbits; the circle is the "perfect" form, and nothing short of perfection can be allowed in the heavens. First comes the Moon; then Mercury, Venus and the Sun, followed by the three other planets then known (Mars, Jupiter and Saturn), beyond which lies the sphere of the so-called fixed stars.

It may sound delightfully simple, but actually it is nothing of the kind, because the theory of planets moving at uniform velocity in circular orbits does not fit the observations – as Ptolemy knew quite well. In particular, the planets do not move steadily from west

The Ptolemaic view of the universe. It was assumed that all celestial orbits must be circular since this is the perfect form. The theory was accepted for more than 1300 years after Ptolemy's time.

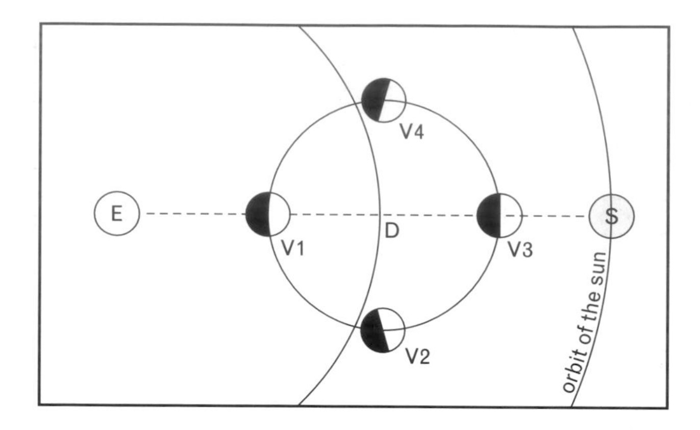

Movements of Venus, according to the Ptolemaic system. E = Earth; S = Sun; V1–V4 =Venus.

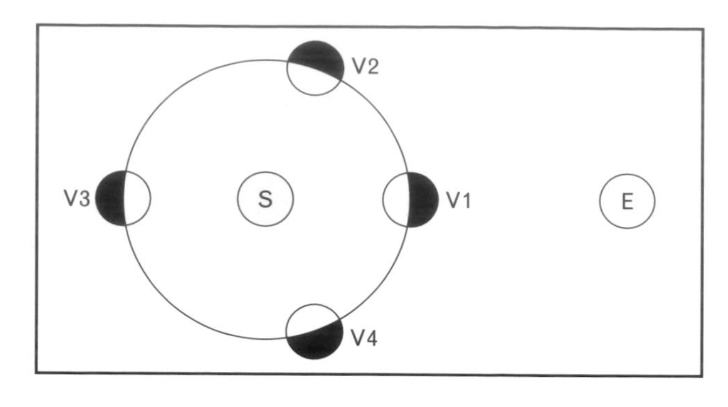

The phases of Venus. VI = Inferior Conjunction; V2, V4 = Elongation; V3 = Superior Conjunction. E = Earth; S = Sun.

to east against the stars. Mars, Jupiter and Saturn regularly stand still and then move in reverse or retrograde direction for a brief period before continuing their eastward march. To overcome this problem, Ptolemy assumed that a planet moved in a small circle or epicycle, the centre of which – the deferent – itself moved round the Earth in a perfect circle. The possibility of non-circular orbits was not considered, so that there really was no other way out of the difficulty. Moreover, a single epicycle would not do; others had to be added, until the whole system had become hopelessly clumsy and artificial.

Mercury and Venus presented problems of their own, and Ptolemy was forced to assume that their deferents remained permanently in a straight line between the Sun and the Earth. At least this explains why Mercury and Venus can never be seen opposite to the Sun in the sky. Centuries later, it was this arrangement which led on to a conclusive disproof of the Ptolemaic theory.

In the schematic diagram showing the movement of Venus in the Ptolemaic system, E represents the Earth at rest in the centre of the universe; S is the Sun, D is the deferent of Venus, and V1 to V4 represent the planet in its four positions as it moves in a small circle or epicycle (obviously the diagram is not to scale). The line EDS must always be straight, and there is one vitally important consequence. Since a planet shines only by reflected sunlight, it can be seen that on the Ptolemaic theory Venus can never be seen as a full disk, or even half. At V1 and V3 the dark hemisphere will be turned towards Earth, so that Venus will be invisible; at V2 and V4 part of the sunlit side will face us, so that Venus will appear as a crescent. Yet as soon as telescopes became available, it was found that Venus showed a whole cycle of phases, from new to full.

So far as we know, the phases of Venus were discovered in 1610 by Galileo, using his newly-built telescope. This was very feeble by modern standards, but it enabled him to make a number of spectacular discoveries: he saw the craters of the Moon, the satellites of Jupiter and the myriads of stars of the Milky Way, but Venus was of special significance. Following his observations, he sent a message to the great mathematician, Johann Kepler: "Haec immatura, a me, iam frustra, leguntur – o.y." In this form the message may be interpreted as "these things not ripe (for disclosure) are read by me" – the letters o.y. are tacked on, since there is no way of fitting them into the original sentence. Rearranged, the letters give: "Cynthiae figuras annulatur Mater Amorum", or "Mother of Loves imitates the phases of Cynthia". The Mother of Loves is, of course, Venus;

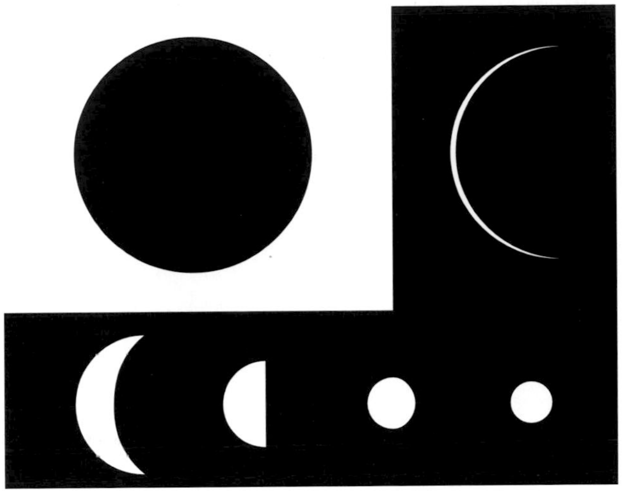

The apparent size of Venus at different phases.

Cynthia is the Moon. Galileo was already a firm believer in the theory that the Earth moves round the Sun instead of vice versa, and the phases of Venus provided conclusive proof that in any event the Ptolemaic system was completely wrong.

The battle had already been joined. In 1543 the Polish cleric Copernicus had published his book *De Revolutionibus Orbium Coelestium,* in which he removed the Earth from its proud position at the centre of the universe and put the Sun there instead; predictably the Church was furious, and "Copernicans" were fiercely persecuted – indeed one of them, Giordano Bruno, was burned at the stake in Rome in 1600 partly because he taught that the Earth moves round the Sun. It is true that Copernicus made many errors, such as suggesting that Venus might be either self-luminous or else transparent, and he was even reduced to bringing back epicycles, but he had taken the first steps. Galileo was not so cautious; he publicized his results and opinions, so that he was brought to trial in Rome and forced into a hollow and completely meaningless recantation. It was not until 1992 that the Vatican finally admitted that Galileo had been correct!

The actual situation in shown in the diagram. Venus is "new" in position V1, half (dichotomy) at V2 and V4; and full at V3. The orbit is almost circular, so that the distance from the Sun does not change much, and on average Venus receives about twice as much solar radiation as does the Earth. The orbit is tilted with respect to that of the Earth; the angle of inclination is 3° 24'. If this were not so, Venus would pass between the Sun and the Earth at each return to "new", and would be seen as a small disk crossing the Sun's face; but the inclination means that these transits do not happen very often. The last was in 1882, while the next will be delayed until 2004.

Position V1 is termed inferior conjunction, while position V3 is superior conjunction – and Venus is virtually out of view, since it is more or less behind the Sun. As the phase increases, the apparent diameter as seen from the Earth shrinks. In the diagram the black circle represents the size at inferior conjunction, when Venus may be no more than 39 million kilometres away (much closer than Mars can ever be), and the last white disk represents the apparent size at superior conjunction, when the diameter has been reduced to less than 10 seconds of arc – smaller than that of remote Saturn. Venus is at its most brilliant during the crescent stage; the increasing phase has to be balanced against the decreasing apparent diameter.

There is an interesting, if minor, problem

connected with the moment of each half-phase, or dichotomy (the word comes from the Greek, and means "cut in half"). Dichotomy should occur at positions V2 and V4, and should be predictable with great accuracy. In fact, dichotomy is always a day or two early when Venus is at eastern elongation, and is therefore a waning crescent in the evening sky; at western elongation, when Venus is waxing in the morning sky, dichotomy is late. The reason is that Venus has a thick, extensive atmosphere which is illuminated by the Sun, giving a spurious effect. The phase discrepancy was first noted in the 1790s, by a skilful and enthusiastic German amateur, Johann Hieronymus Schröter. In a paper published years ago by one of the present authors (PM) it was christened the "Schröter effect", and the term has now come into general usage.

When Venus is in the crescent stage, very keen-sighted people can make out the phase with the naked eye; binoculars show it very clearly, but even in a powerful telecope, it is frankly unrewarding; all that can be seen is the top of a thick layer of clouds, and there are no definite markings, though on most occasions it is possible to make out a few vague shadings. It is therefore not surprising that until recently we have had very little idea of what Venus is really like. As recently as 1960 it was still being referred to as "the planet of mystery"; our first reliable information did not come until after the onset of the Space Age.

VENUS THROUGH THE TELESCOPE

2

The first telescopic drawings of Venus were made in 1645, by Fontana; in 1725 Bianchini even produced a map of the planet. However, the markings recorded by these early observers were certainly spurious. G. D. Cassini drew the planet in the 17th century, and in the following century better observations were obtained by William Herschel and Johann Schröter; however, the real nature of Venus remained unknown

Drawings of Venus by G. D. Cassini.

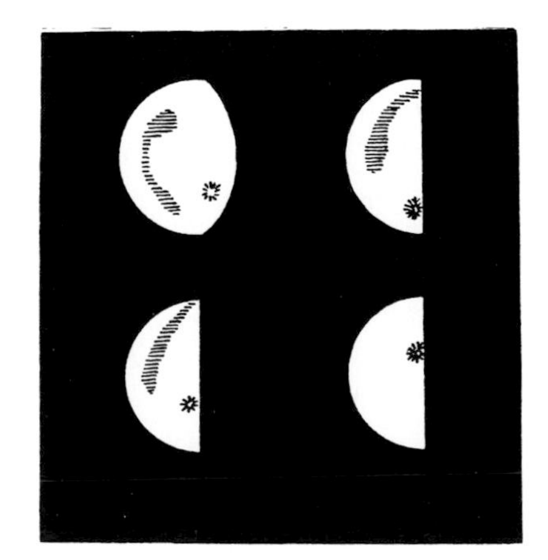

The first attempt at recording surface details on Venus was made in 1645 by a Neapolitan lawyer and amateur astronomer, Francesco Fontana. He recorded "a dark patch in the centre of the disk" of the planet. Fontana's telescope was home-made, and probably not much better than Galileo's; earlier, in 1636 and 1638, he had sketched Mars, showing a circular disk with a ring inside it and a central dark patch. There is no doubt that these effects, with both Mars and Venus, were purely optical.

Something which seemed to be more positive came in 1667, with the work of Giovanni Cassini, who at that time was observing from Bologna. His first sketch was made on 14 October 1666, and for the next few months he undertook a whole series, showing various bright and dusky patches; from these he deduced the first estimated rotation period – 23h 21m, only a little shorter than that of the Earth. Subsequently he left Italy to become the first Director of the newly-formed Paris Observatory, and from the less transparent skies of France he was unable to recover the shadings of Venus – which makes it look doubtful whether his original observations were anything but spurious. This is not to belittle him; from Paris he discovered four of Saturn's satellites (Iapetus, Rhea, Dione and Tethys) and he also found the main division of the Saturnian ring system, so that he was undoubtedly an expert observer.

The first "map" was drawn by Francesco Bianchini, from 1726; he observed from Rome – often from the grounds of the Vatican – with a 6.4-cm refractor with a focal length of 20 metres. His account of his work is truly riveting, and the results were spectacular inasmuch as he recorded what he believed to be oceans and continents; he even

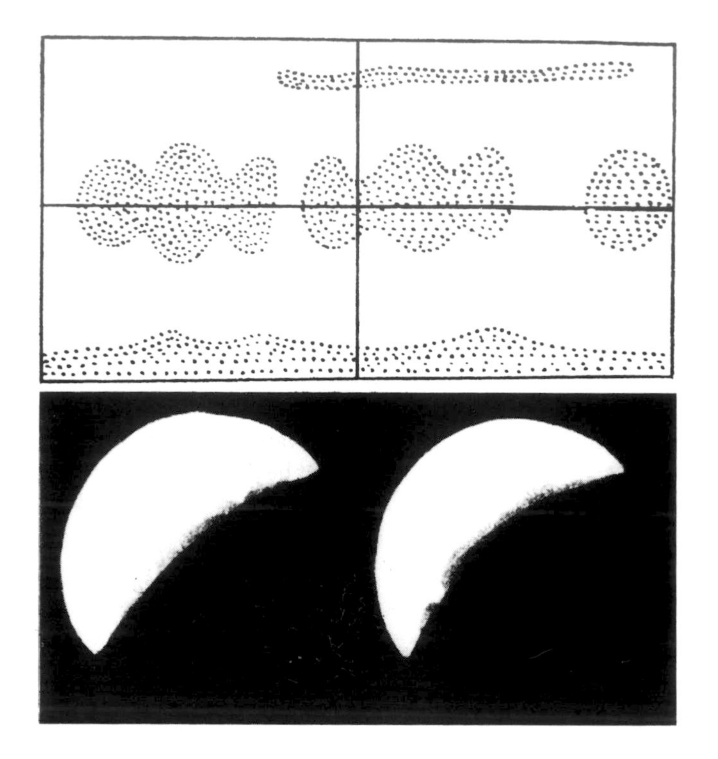

Observations of Venus by Bianchini. (Above) **Map of the surface.** (Below) **drawings of the planet. From *Hesperi et Phosphori Nova Phaenomena*, Rome, 1727.**

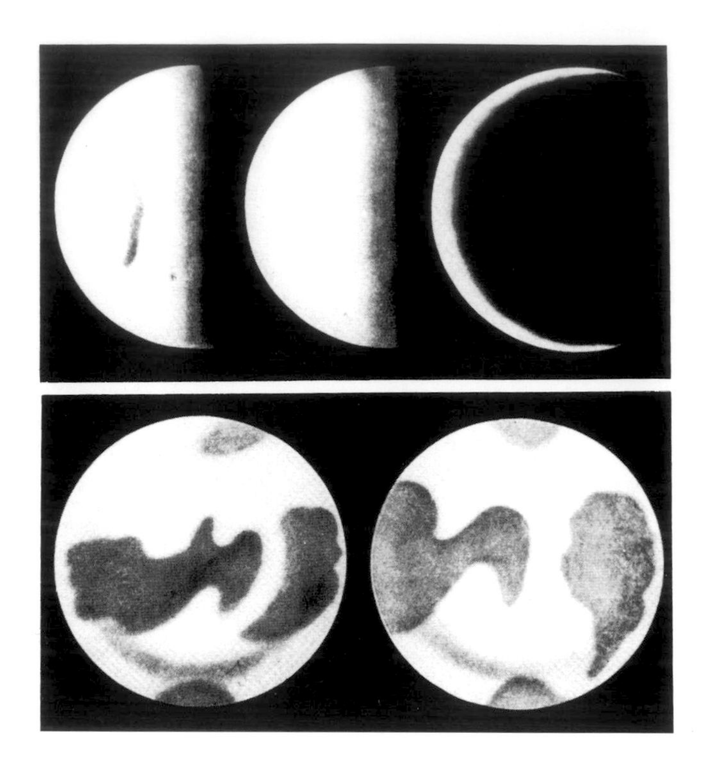

Venus as drawn by Johann Schröter.

gave them grandiose names, and deduced a rotation period of 24d 8h. Yet once again we have to dismiss his features as purely optical; apart from Galileo's discovery of the phases the first really useful telescopic result was due to the Russian astronomer M. V. Lomonosov, in 1761. In that year Venus passed in transit across the face of the Sun, and Lomonosov found that Venus "is surrounded by a considerable atmosphere equal to, if not greater than, that which envelops our earthly sphere".

The two main planetary observers of the late eighteenth century were William Herschel and Johann Hieronymus Schröter. Herschel was, of course, one of the greatest observers of all time; in 1781 he discovered the planet Uranus, and during his long career he recorded hundreds of new double stars, clusters and nebulae. He was also the first to give a reasonable picture of the shape of the Galaxy. He made his own telescopes, which were much the best of their time; the largest had a focal length of 40 feet (19 m) and a mirror 49 inches (125 cm) in diameter, though most of his best work was done with smaller instruments.

Schröter was an amateur who set up his observatory at Lilienthal, near Bremen (where he was chief magistrate) and equipped it with the best telescopes he could obtain – including one of Herschel's. He systematically observed the Moon and planets from 1778 until 1814, when his observatory was destroyed by the invading French army and his brass-tubed telescopes looted by the soldiers – who mistook them for gold. He was the first really great observer of the Moon, and he also drew good maps of Mars, though admittedly he did misinterpret them, believing the features to be atmospheric. He was also involved in the search for a new planet between the orbits of Mars and Jupiter; indeed the third asteroid,

Juno, was found from Lilienthal by his assistant, Karl Harding. Schröter was not a good draughtsman, but he was a painstaking observer who seldom made a really serious error.

He did not find Venus an easy object. Between 1779 and 1788 he failed to detect any markings at all, but on 28 February 1788 he "perceived the ordinarily uniform brightness of the disk to be marbled by a filmy streak" and subsequently he saw other markings, though all were diffuse and ill defined. With Venus his interpretation was correct, and he realized that what he was seeing was the top of a dense, cloudy atmosphere.

Moreover, he found that the light of the disk falls away perceptibly toward the terminator, which indicates absorption in the planet's atmosphere; also, the horns of the crescent are seen to be prolonged beyond the semi-circle, which again could not happen in the absence of an atmospheric mantle. He discovered the phase anomaly, which is now known as the Schröter effect, and he did his best to derive a rotation period. His final value was 23h 21m 7s.977. It may sound peculiar to give a period to an accuracy of less than a second, but it is only fair to add that Schröter was well aware of the difficulties, and he commented:

'The circumstances that there are seen on this planet none of the flat spherical forms as are conspicuous on Jupiter and Saturn, none of the strips of longitudinal spots parallel to the equator which are seen on those planets … and which point out a certain stretch of atmosphere, give room to infer that Venus … performs its rotation round its axis in a much longer space of time than these planets … and this is actually confirmed by my observations of the distinct part of Venus.'

He gave the axial inclination as 15 degrees to the perpendicular, as against 23 degrees for the Earth.

Herschel's observations of Venus were less systematic, but he did look at the planet from time to time, and one of his records, for 19 June 1780, is worth quoting:

'There is on Venus a bluish, darkish spot, and another, which is rather bright … continued observations were made of these and other faint spots. The instrument used was a 20-foot Newtonian reflector, furnished with no less than 5 different object specula, some of which were in the highest perfection of figure and polish; the power generally 300 and 450. But the result of them would not give me the time of the rotation of Venus. For the spots assumed often the appearances of optical deceptions, such as might arise from prismatic affections; and I was unwilling to lay any stress upon the motion of spots, that either were extremely faint and changeable, or whose situation could not be precisely ascertained. However, that Venus has a motion on its axis cannot be doubted, from these observations; and that she has an atmosphere is evident, from the changes I took notice of, which surely cannot be on the solid body of the planet.'

Herschel and Schröter never met face to face, but they corresponded frequently, and were on excellent terms. Curiously, their one major disagreement was with regard to Venus. On 28 December 1789, using magnifications of between 161 to 370 on his 7-foot focus Herschel telescope, Schröter saw that the planet's southern cusp was blunted, while beyond it there was a small luminous speck. He saw it again in 1790 and 1791 and became convinced that it was a very lofty "enlightened mountain" catching rays of the Sun – as does often happen with

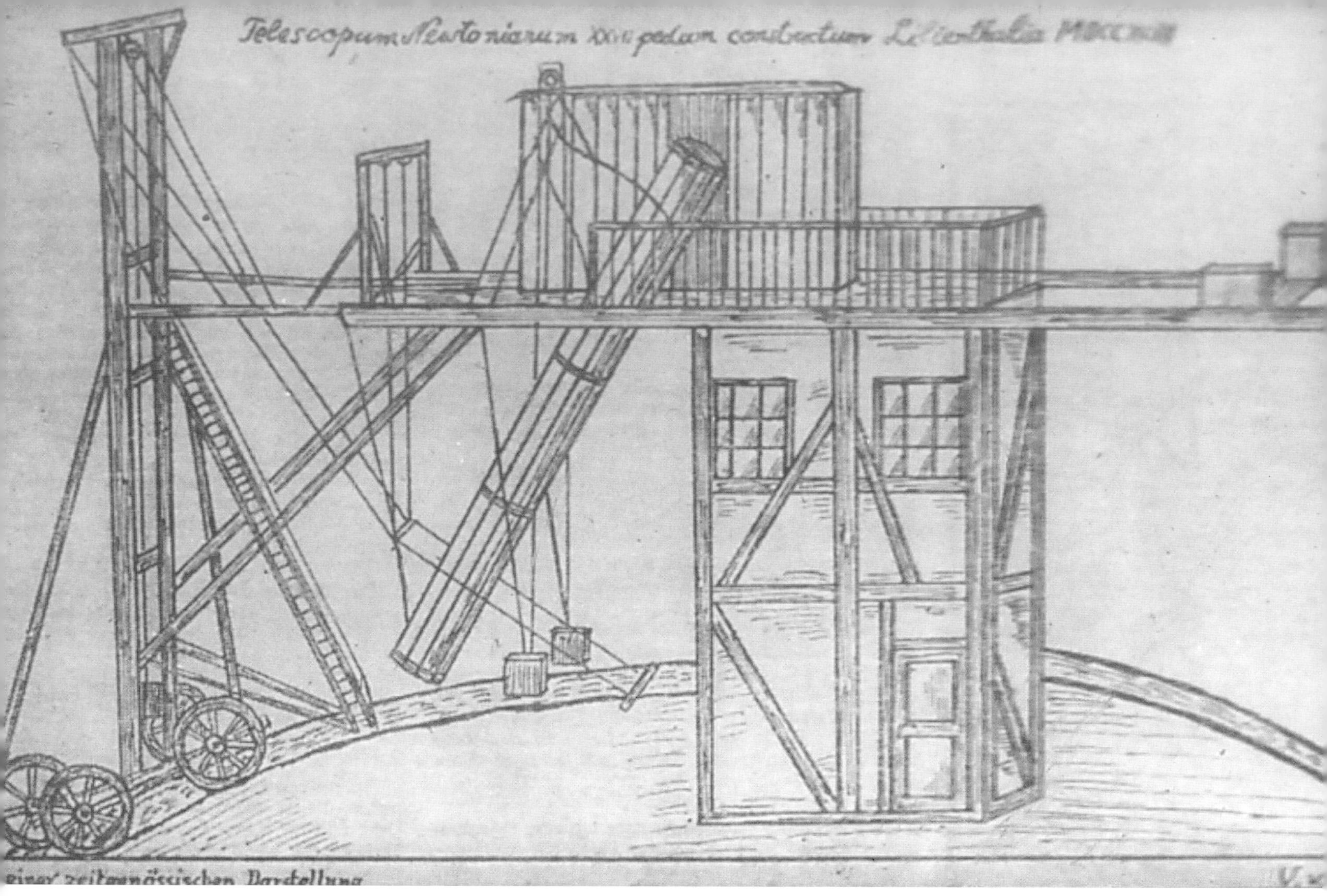

Schröter's observatory at Lilienthal. He was the first great observer of the Moon

the Moon, where peaks near sunrise or sunset can often be seen apparently well clear of the terminator. Eventually he wrote:

'Though we cannot suppose a smaller, but rather a greater force of gravity on the surface of Venus than our own globe, Nature seems, however, to have raised on the former such great inequalities, and mountains of such enormous height, as to exceed four, five or even six times the perpendicular elevations of Cimboraco, the highest of our mountains.'

The estimated height of Venus's atmosphere was given as from 4 to 5 kilometres.

During the latter part of the nineteenth century Schröter's mountain theory was very much alive. One strong supporter of it was Etienne Trouvelot, a Belgian observer who spent some time first in America and then at the Meudon Observatory near Paris. Like Schröter, Trouvelot recorded luminous spots beyond the planet's terminator, and he also drew the familiar "polar hoods", which have often been called polar caps – though at the time there was no proof that they did indeed mark the poles of Venus. It was suggested that there might be vast, elevated polar plateaux, and in 1878 Trouvelot wrote that

'the polar patches are distinctly visible, the southern one being more brilliant. Their surface is irregular, and seems like a confused mass of luminous points, separated by comparatively sombre intervening spaces. This surface is undoubtedly very broken, and resembles that of a mountainous district studded with numerous peaks ... The polar spots seem to be bristling with peaks and needles.'

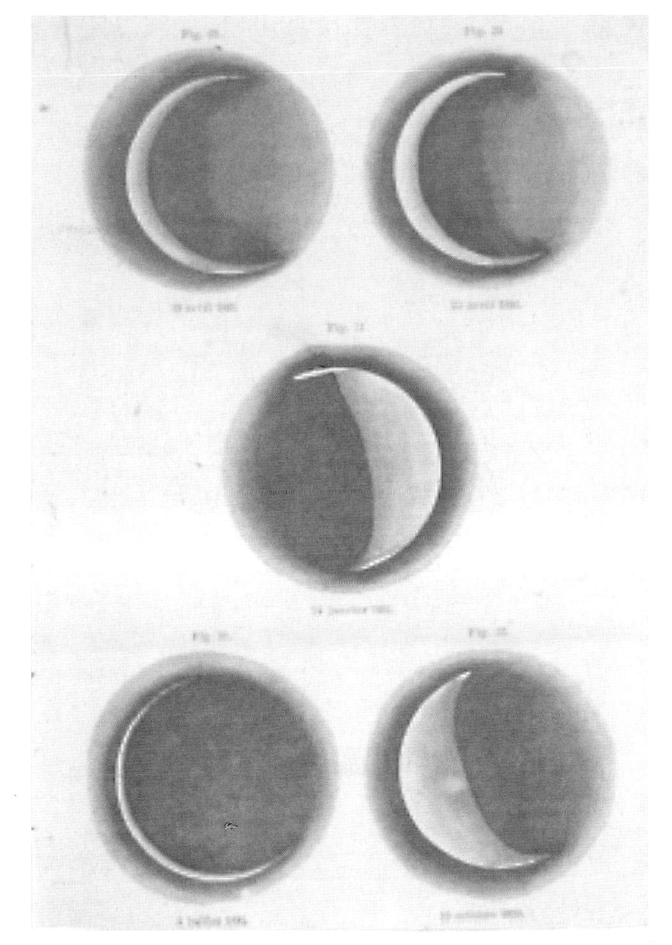
Drawings of Venus by Etienne Trouvelot.

Venus, as drawn by William Herschel in 1780.

Schröter's "Enlightened Mountain".

The polar hoods are often very much in evidence, and can be striking. We now know that they are indeed polar, but of course they are purely cloud phenomena – and though the surface of Venus is extremely mountainous in places, the volcanoes and peaks there bear no relation to the phenomena reported by pre-Space Age observers.

Meanwhile, continued efforts were made to measure the rotation period of Venus – purely by observations of the surface markings, vague and transient though they were. Generally speaking the values given were of the order of 24 hours, and some of them were taken to an absurd degree of accuracy. At Lussinpiccolo, in 1895, Leo Brenner gave a value of 23h 57m 36s.2396, amending it in the following year to 23h 57m 36s.27728. One cannot help feeling that this is rather like estimating the age of the Earth to the nearest minute.

Then, however, G. V. Schiaparelli, the Italian astronomer who will always be remembered for his observations of the "canals" of Mars, produced an astronomical bombshell in the form of a rotation period of 224d 16h 48m, which is the same as the time taken for Venus to complete one orbit of the Sun. In other words, Schiaparelli came to the conclusion that Venus keeps the same hemisphere turned permanently sunward.

Schiaparelli was undoubtedly a skilful observer, even though he was so completely wrong in drawing hard, sharp, even double linear features on Mars. His main observations of Venus were carried out from Milan in 1877–8 and in 1892; sensibly, he carried out his work when Venus was high above the horizon – the fact that the Sun was also well in view was a real handicap, but there was no real alternative; when Venus is a brilliant object against a reasonably dark background, it is bound to be low down, and the seeing conditions will be poor.

From studies of the south polar hood, Schiaparelli derived what he believed to be a definite rotation period. In fact there would have been nothing particularly surprising about it. The Moon behaves in this way with respect to the Earth, and so do the main satellites of the other planets relative to their primaries. Tidal friction over the ages is responsible, and the idea of Venus having a similarly captured rotation was perfectly reasonable. Schiaparelli also studied the delicate surface markings of Mercury, and came to a similar conclusion – a rotation period equal to the revolution period (in the case of Mercury, 88 Earth-days). His opinions carried a great deal of weight, and for Mercury the synchronous rotation period was still accepted until radar work in the 1960s showed it to be wrong (the true value is 58.6 Earth-days, or two-thirds of a Mercurian year). Venus was more of a problem; some observers agreed with Schiaparelli, while others did not. It is, incidentally, worth noting that as late as 1955 the synchronous period was still accepted by Audouin Dollfus, one of the leading modern observers of the planets.

It is interesting to look back at the various rotation periods proposed for Venus before the problem was finally solved, in 1962, by radar methods. Estimates ranged between less than 24 hours and up to 225 days. The one thing that was *not* suspected was that the period might be longer than the Venusian year – and yet this proved to be the truth. One of us (PM) published a monograph in which all the rotation period estimates were listed. Before 1962 there were over ninety of them – and all turned out to be wrong. A selected list is given in Appendix 4.

Let us now come to the spoke-like markings of Venus, sometimes nicknamed the "Venusian canals". They do not exist, and they were never regarded as artificial, but the story is worth examining, and inevitably it is to some extent linked with the notorious canals of Mars.

The Martian network of canals was first reported in detail by Schiaparelli in 1877. They were described as being unnaturally regular, forming a planet-wide réseau, and subject to doubling or "gemination". So far as their origin was concerned, Schiaparelli kept an open mind; his own name for them was "channels" (*canali*), though it was naturally translated as "canals", and Schiaparelli was careful not to dismiss the idea that they might be artificial. He certainly believed them to be channels through which water flowed equatorward from the shrinking polar ice-caps during spring and early summer on Mars. They were confirmed in 1886 by Perrotin and Thollon, at the Observatory of Nice, and from then onward they became thoroughly fashionable. Not until the Mariner results of the 1960s were they finally banished to the realms of myth.

Schiaparelli was followed by Percival Lowell, who founded the observatory at Flagstaff in Arizona principally to observe Mars. Between 1894 and his death in 1916 he and all his colleagues recorded hundreds of Martian canals, and attributed them to channels built by intelligent engineers in an attempt to defeat the growing water shortage on Mars by means of a vast irrigation system.

Lowell's conclusions, as well as his observations, were hotly challenged. Other

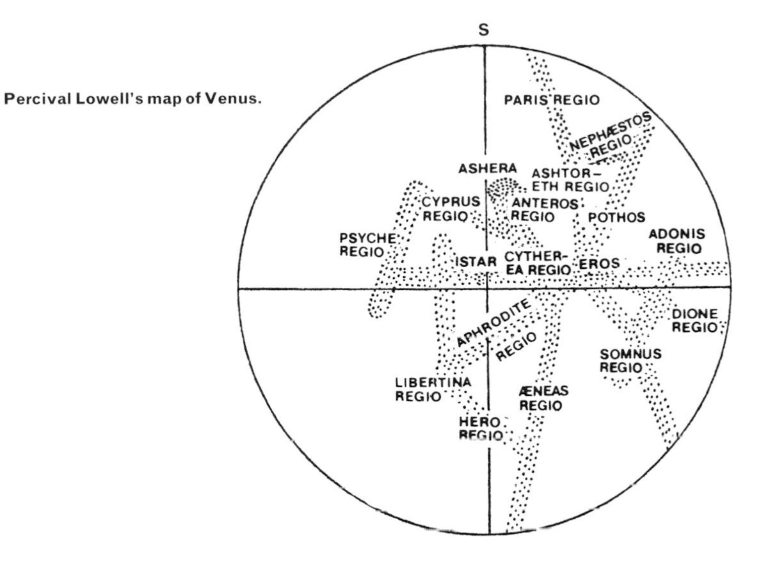

Percival Lowell's map of Venus.

observers, using telescopes as powerful as Lowell's 61-cm refractor, either failed to see canals at all, or else recorded them as faint, irregular streaks. Moreover, Lowell showed straight lines not only on Mars, but on other bodies such as Mercury, the satellites of Jupiter – and Venus.

Vague streaks on Venus had been recorded by Perrotin from Nice, but Lowell's chart of the planet, published in 1897 with an accompanying description, was much more definite. He even named the features he believed he had seen. From a dark patch, Eros – a sort of focal centre – he drew dark, well-defined strips which he named Adonis Regio, Aeneas Regio and so on. He was quite convinced that the markings were unchanging, and that the synchronous rotation period was correct, so that the same hemisphere was permanently sunlit. His description of the markings runs as follows:

'The markings themselves are long and narrow but, unlike the finer markings of Mars, they have the appearance of being natural, not artificial. They are not only permanent, but permanently visible whenever our own atmospheric conditions are not so poor as to obliterate all detail in the disk. They are thus evidently not cloud-hidden at any time ... There is no distinctive colour in any part of the planet other than its general brilliant straw-coloured hue. The markings, which are of a straw-coloured grey, bear the look of being ground or rock, and it is presumable from this that we see simply barren rock or sand weathered by aeons of exposure to the Sun. The markings are perfectly distinct and unmistakable, and conclusive as to the planet's period of rotation. There is no certain evidence of any polar caps.'

Lowell's chief critic was Eugenios Antoniadi, a Greek-born astronomer who spent most of his life in France and made extensive use of the great 83-cm refractor at the Meudon Observatory. Antoniadi was possibly the best of all planetary observers of the pre-Space Age period, and he had no patience with canals on Mars, Venus or anywhere else. In 1897 he gave his views very frankly:

'I refrain from discussing the aphroditographic work of most of our contemporaries who, forgetting that Venus is decidedly clad in a dense atmospheric mantle, cover what they call the "surface" of the unfortunate planet with the fashionable canal network, dividing it into clumsy melon slices having their radii now on the cusps and then on the visual ray. Such discussion would lead to some curious conclusions*

*The hideous adjective "aphroditographic" is perfectly correct. One can only hope that it will never come into general use.

The Ashen Light. The origin of this phenomenon is still not understood. Drawing by Patrick Moore, 15-in (38-cm) reflector. The Ashen Light is greatly exaggerated for the sake of clarity.

in which we should not be long in finding that Venus does not constantly present the same face to the Sun – but to the Earth.'

For obvious reasons, the streaks of Venus were never recorded photographically; one of the keenest-eyed of all observers, Edwin Emerson Barnard, looked for them with the world's largest refractor (the Yerkes 101-cm) with a total lack of success, and gradually belief in their existence faded away. Yet it was slow to die. Occasionally it was even suggested that the streaks might be best seen with small telescopes (!) and in the 1950s drawings showing a dark central patch and radiating streaks were published by R. M. Baum, using a 7.5-cm refractor and a 15.2-cm reflector, but once again there is no doubt that these effects were purely optical. The Lowell-type markings on Venus are just as unreal as the Martian canals.

Before leaving the nineteenth century, there are three more topics of historical interest: the phantom satellite, the Ashen Light, and the rare but fascinating transits.

As we have noted, G. D. Cassini, from Paris, discovered four of Saturn's satellites. Then, in 1686, he made what was thought to be an equally important discovery. On 18 August, using a telescope of 34-feet (10m) focal length, he observed Venus and "saw at a distance of 3/5 her diameter, eastward, a luminous appearance, of a shape that seemed to have the same phase with Venus, which was then gibbous on the western side. The diameter of this object was nearly one-quarter that of Venus." He made other similar observations, and became convinced of the reality of a satellite. On 23 October 1740 the famous telescope-maker James Short also recorded an attendant body with the same phase of Venus itself, and there were other sporadic reports until 1764, after which the satellite disappeared from the observation books. A brilliant object such as Venus is very prone to produce "telescopic ghosts", and if Venus had a satellite of any appreciable size it would most certainly have been detected by now. Per Wargentin, a famous Swedish observer, commented that he owned a telescope which never failed to show companions to Venus or any other really bright object!

The Ashen Light, however, falls into a very different category. This is the name given to the dim visibility of the non-sunlit side of the planet when Venus is in the crescent stage. Of course, the same phenomenon is easy to see with the Moon, and has been known for centuries; it is known to country folk as "the Old Moon in the Young Moon's arms", and there is absolutely no mystery about it. It is due simply to light reflected on to the Moon from the Earth, so that it is generally termed the Earthshine. But it is not nearly so easy to explain a similar appearance with Venus.

Apparently the Ashen Light was first reported in 1643 by Giovanni Riccioli, a Jesuit priest at Bologna, whose chief claim to fame is that he drew up a map of the Moon and gave the main craters the names which are still in use today. Other observations of the same type followed, and in fact the phenomenon has been recorded by almost every observer of the planet at one time or another. It is not easy to see, and the best method is to block out the bright crescent by means of an occulting bar, or some similar device, to guard against effects of contrast. Both the present authors have recorded it frequently, and have seen it as so prominent that it cannot be dismissed as a contrast effect.

Inevitably there were some curious theories about its origin – and it is worth quoting Franz von Paula Gruithuisen, an enthusiastic German of the mid-nineteenth century, who was in no doubt about it. He pointed out that the Light had been seen in 1759 and again in 1806, an interval of 47 terrestrial days or 76 Venusian years, and wrote:

'We assume that some (Venusian) Alexander or Napoleon then attained universal power. If we estimate that the ordinary life of an inhabitant of Venus lasts 130 Venusian years, which amounts to 80 Earth years, the reign of an Emperor of Venus might well last for 76 Venusian years. The observed appearance is evidently the result of general festival illumination in honour of the ascension of a new emperor to the throne of the planet.'

Later on, Gruithuisen modified his theory. Instead of a Venusian Coronation, he suggested that the Light might be due solely to the burning of large stretches of jungle to produce new farm land, and added that "large migrations of people would be prevented, so that possible wars would be avoided by abolishing the reason for them. Thus the race would be kept united."

There are perhaps certain objections to these theories, and others were proposed – for example phosphorescent oceans; this was supported by no less a person than William Herschel, who saw the Light several times around 1790. In 1872 P. de Heen proposed that the Light might be due to Venusian equivalents of terrestrial aurorae, and it now seems that unless the cause really is due to sheer contrast – which is hard to believe – some sort of electrical phenomenon must be responsible.

Transits of Venus across the face of the Sun do not happen often. Were the orbits of Venus and the Earth in the same plane, a transit would occur at every inferior conjunction, but this is not the case. The orbit of Venus is inclined at an angle of 3°.4, so that at most inferior conjunctions the planet passes either above or below the plane of the Sun in the sky. At the present epoch transits are seen in pairs, the components of a pair being separated by eight years, after which no more transits are seen for over a century. Thus there were transits in 1631, 1639, 1761, 1769, 1874 and 1882; the next (Appendix 3) will be on 8 June 2004 and then 5/6 June 2012, after which we must wait until 11 December 2117.

Obviously, Mercury and Venus are the only planets which can pass in transit (not counting various small asteroids, which are much too small to be seen when crossing the Sun's face). In 1627 the great mathematician Johannes Kepler finished what was destined to be his last work, a set of new and more accurate tables of planetary movements which he called the Rudolphine Tables in honour of his old benefactor, the Holy Roman Emperor, Rudolph II. He was able to show that both Mercury and Venus would transit the Sun in 1631 – Mercury on 7 November and Venus on 6 December. By 1631 Kepler was dead, but the transit of Mercury was successfully observed by the French astronomer Pierre Gassendi.

Encouraged by this success, Gassendi naturally expected to be equally fortunate with the transit of Venus, since Venus is not only much closer to us than Mercury but is also much larger. He left nothing to chance. Fearful that Kepler's prediction might be in error, he began watching the Sun on 4 December, and went on observing during the 6th and 7th. To his surprise, and disappointment, he saw nothing. The reason is now known; the transit did occur as Kepler predicted, but it took place during the northern night of 6–7 December, when the Sun was below the horizon from France.

Kepler had predicted no more transits before 1761, but fresh calculations were made by a young English amateur, Jeremiah Horrocks, showing that a transit would occur on 24 November 1639 (Old Style; the New Style date is 4 December). Horrocks' calculations were not finished until shortly before the transit was due, and he had time only to inform his brother Jonas, near Liverpool, and his friend William Crabtree, at Manchester.

Edmond Halley, from a contemporary painting.

From Hoole, Horrocks began to watch the Sun on 23 November. On the following day he was unable to begin observing before 3.15 in the afternoon, but then, in his own words:

'At this time an opening in the clouds, which rendered the Sun distinctly visible, seemed as if Divine Providence encouraged my aspirations; when, O most gratifying spectacle! the object of so many earnest wishes, I perceived a new spot of unusual magnitude, and of a perfectly round form, that has just wholly entered upon the left limb of the Sun, so that the margin of the Sun and spot coincided with each other, forming the angle of contact.'

He followed the planet until after sunset, an hour and a half later, and was able to make some useful measurements. Jonas was clouded out; Crabtree had bad luck with the weather, but did manage to see Venus just before sunset, when the clouds broke up for a few seconds.

This was the first definite observation of a transit of Venus. It is true that Arab astronomers may have seen one as long ago as the year 839, but we cannot be sure; what they observed could so easily have been a naked-eye sunspot. In any case, Horrocks' prediction was a brilliant piece of work. Had he lived he would have gone on to great things, but unfortunately he died in 1641 at the early age of twenty-two.

Before the 1761 transit took place, Edmond Halley realized that these transits might be used to measure the length of the astronomical unit, or distance between the Earth and the Sun. Since our world is a globe, and not a point, the position of Venus against the Sun, and hence the time of entry on to and departure from the solar disk, will not be the same for different observing stations; and by careful timing, information can be gained which will provide the essential clue. As the entire method is now completely obsolete there is no point in describing it further, but at all events the transits of 1761 and 1769 were carefully studied at many observing sites. Results from the 1761 transit were unsatisfactory, apart from Lomonosov's detection of the atmosphere of Venus, and the derived value of the astronomical unit could only be given as somewhere between 160,000,000 km and 129,000,000 km. However, the method was sound enough, and extensive preparations were made to observe the transit of 1769.

The main trouble was that it was impossible to give an accurate timing for the start of the transit, owing to an effect known as the Black Drop. As Venus passes on to the Sun's disk, it seems to draw a strip of blackness

Photograph of the 1874 transit of Venus.

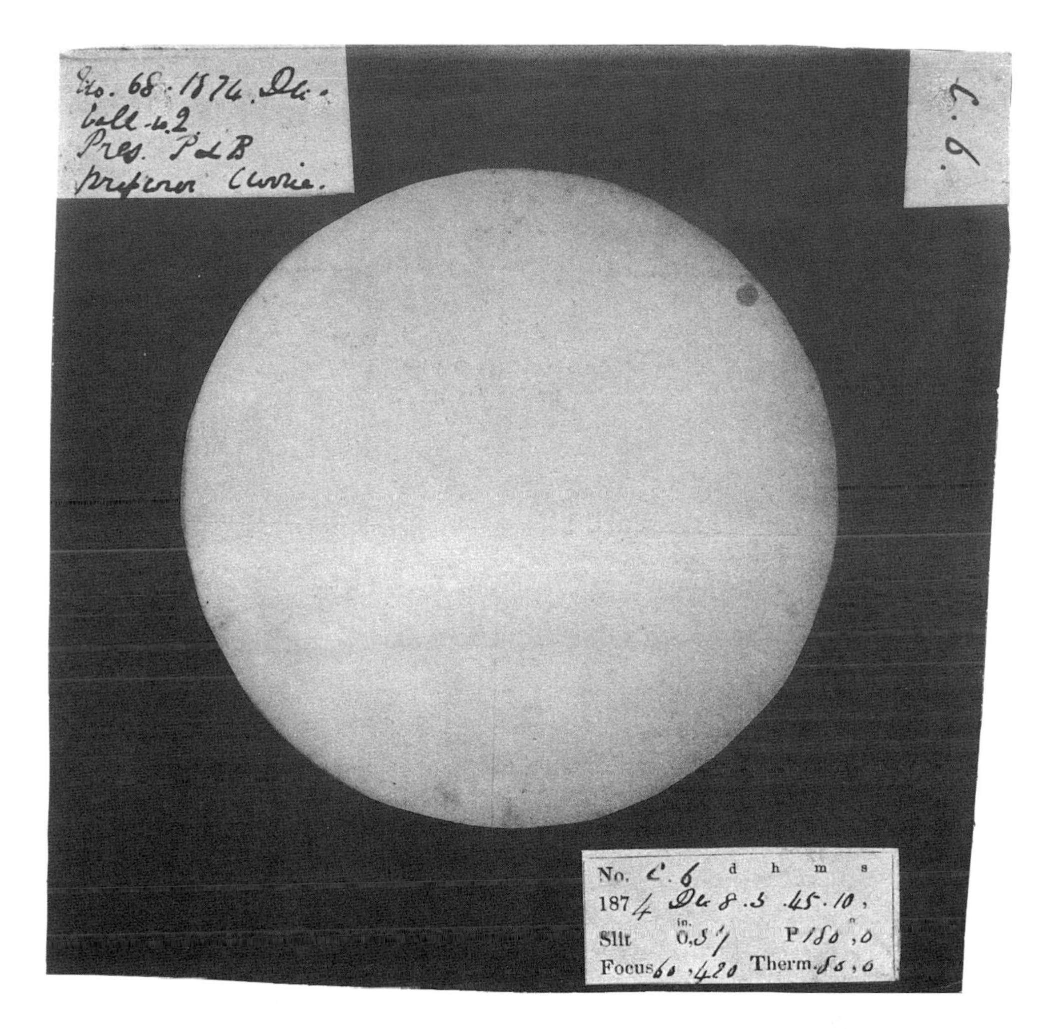

after it, and when this disappears the transit has already begun. Of course, this effect is due solely to Venus's atmosphere, but there is very little that can be done about it.

Despite the Black Drop effect, astronomers did not intend to miss the opportunity, and the English Astronomer Royal, Nevil Maskelyne, was particularly enthusiastic. The best observing site was expected to be in the South Seas, and with the express approval of King George III a ship was fitted out for the journey and put under the command of Captain James Cook; the senior astronomer was Charles Green, who had been an assistant at Greenwich Observatory and was well known to Maskelyne. After some deliberation the Admiralty settled on Tahiti, and on 26 August the ship *Endeavour* sailed from Plymouth. By 10 April it was standing off Tahiti; a temporary observatory was set up, and there were no mishaps, apart from the fact that an essential quadrant was stolen by one of the local natives and was recovered only with some difficulty. On the day of the transit the weather was perfect, and observations were made by Cook, Green, and Daniel Solander. Cook's own account reads as follows:

'The first appearance of Venus on the Sun was certainly only the penumbra, and the contact of the limbs did not occur until several seconds after … it appeared to be very difficult to judge precisely of the times that the internal contacts of the body of Venus happened, by reason of the darkness of the penumbra of the Sun's limb, it being there nearly, if not quite, as dark as the planet. At this time a faint light, much weaker than the rest of the penumbra, appeared to converge

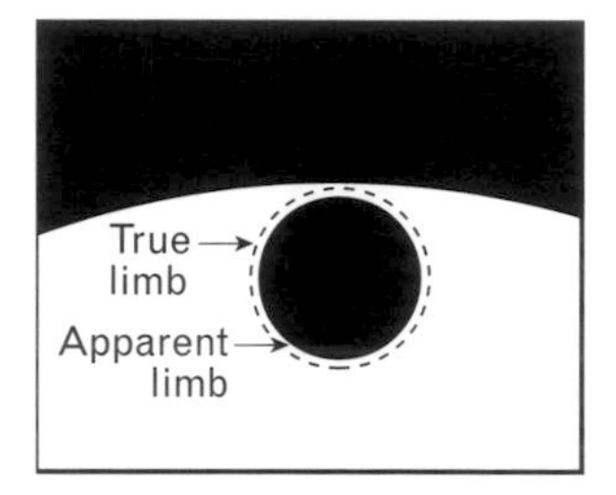

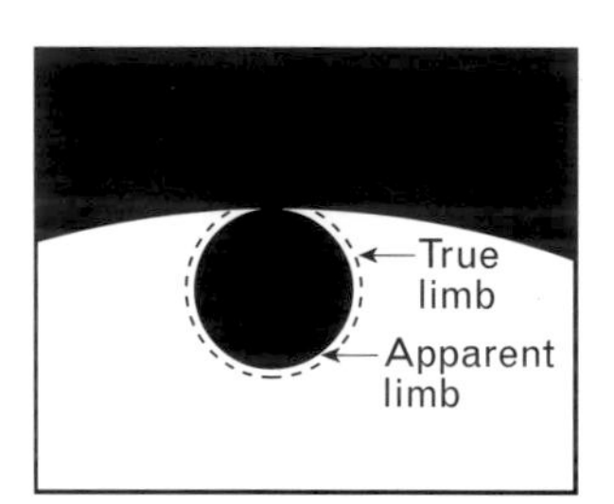

The Black Drop, an effect produced by the dense atmosphere of Venus.

toward the point of contact, but did not quite reach it. This was seen by myself and the other two observers, and was of great assistance to us in judging of the time of internal contacts of the dark body of Venus, with the Sun's limb ... I judged that the penumbra was in contact with the Sun's limb 10" sooner than the time set down above; in like manner at the egress the thread of light was wholly broke by the penumbra ... The breadth of the penumbra appeared to me to be nearly equal to ⅛ of Venus semi-diameter.'

Cook gave the apparent diameter of Venus as 56 seconds of arc (the maximum apparent diameter as seen from Earth is in fact about 66 seconds). The *Endeavour* left Tahiti on 13 July, and continued on its voyage of exploration which is part of our history. The real tragedy was that when the ship was near Java, Charles Green died.

The transit was observed from eighty stations in all, and there were 150 sets of measurements. On the whole, the results were better than those of 1761. They were carefully analysed by the Finnish mathematician Anders Lexell, who gave a value for the solar parallax of 8".63, corresponding to an astronomical unit of 153,000,000 km; later Simon Newcomb revised this to 8".79, which is very near the truth. Yet it was painfully clear that because of the Black Drop, the whole method was much less precise than had been expected.

Historical references to the 1761 and 1769 transits would be incomplete without describing the incredible series of misfortunes suffered by a French astronomer, Guillaume Legentil. Knowing that the 1761 transit would be favourably seen from India, he set out in the previous year. Originally he meant to go to Rodriguez, but decided to change his site to Pondicherry, and made his way there on a French frigate. Unfortunately one of the usual wars between France and England was in progress, and about this time Pondicherry fell to the English, so that Legentil had to turn back. Before he could reach land the transit was over, and all he could do was to make some rough observations from the ship, using improvised equipment. Rather than risk a second delay he elected to wait in India for the next eight years, and observe the 1769 transit instead. Again he altered his site; again he was unlucky. The transit occurred on 3 June 1769. 1 and 2 June were glorious days, but the 3rd was hopelessly cloudy, and Legentil saw nothing at all, though his companions who had remained behind at the original site had a perfect view. It was rather too long to wait for the next transit (that of 1874), and accordingly Legentil packed what belongings he could and set off for home. Twice he was shipwrecked, and eventually reached France after a total absence of eleven years – to find that he had been presumed dead, and that his heirs were preparing to distribute his property...

VENUS BEFORE THE SPACE AGE

3

Little really new information was obtained until the development of photography and, in particular, spectroscopy. Spectroscopic analysis showed that the atmosphere consisted largely of carbon dioxide, so that the surface was likely to be very hot indeed, but the existence of oceans – and even life – was certainly not ruled out.

It cannot honestly be said that a great deal of progress in our knowledge of Venus was made during the first half of the twentieth century, apart from spectroscopic observations of the upper atmosphere. Telescopic work was continued, sometimes with very large instruments. E. M. Antoniadi made extensive use of the 83-cm refractor at the Meudon Observatory, Paris, between 1900 and 1939, but saw nothing definite; with the same telescope, one of us (PM) has been similarly unsuccessful since 1950. In 1956 G. P. Kuiper used the 208-cm reflector at the McDonald Observatory in Texas, but recorded only a suspicion of a fine-scale mottling over the entire disk. Frequent observations were also made at the high-altitude Pic du Midi Observatory in the Pyrenees by Audouin Dollfus and his colleagues, and various "maps" were produced, but all in all it seemed rather pointless to try and give permanent representations of cloudy structures which were so obviously transitory.

Of course, there were many reports from observers using smaller telescopes; thus in 1924 W. H. Steavenson, one of the few amateurs of modern times to have served as President of the Royal Astronomical Society, used a power of 280 on a 15.2-cm refractor to record "a marking which was much more prominent than any I have seen before. Its most conspicuous portion took the form of a dusky band stretching westwards towards the limb just south of what would be the centre of the true disk ... it should be visible with any telescope over 7.6-cm aperture." From this and other features, Steavenson derived a rotation period of 8 days.

All this was highly nebulous, and neither was the inclination of the axis known; estimates ranged from as low as 5 degrees up to the perpendicular, so at least there was plenty of variety!

Photographic results were a little more helpful. In 1924 W. H.Wright carried out pioneer experiments with filters, and then, in 1927, an excellent series of images was obtained by F. E. Ross at Mount Wilson, using the 152-cm and 253-cm reflectors. The most distinct features were seen at ultraviolet wavelengths, confirming that they related to the upper cloud layers rather than to any solid surface. It is very unfortunate that the original Ross plates have been lost.

Photographs taken during the early 1950s by Kuiper at the McDonald Observatory and by N. A. Kozyrev in Russia were similar, and R. S. Richardson used the Mount Wilson reflector to good effect. He deduced that the rotation period must be long, and he also

The 83-cm Meudon refractor. This magnificent instrument was used by E. M. Antoniadi between 1900 and 1939. He saw no definite markings on Venus. The observer here is Audouin Dollfus.

seems to have been the first to suggest that the rotation might be in the retrograde sense – that is to say east to west, opposite to that of the Earth and Mars.

However, the most important work was spectroscopic. Earlier attempts at analyzing the planet's upper atmosphere had been made now and then, and the presence of water vapour had been suspected, but these results subsequently proved spurious. The breakthrough came in 1932, due to work by W. S. Adams and T. Dunham at Mount Wilson, using equipment fitted to the 254-cm (100-inch) Hooker reflector – then not only the most powerful telescope in the world, but in a class of its own. Adams and Dunham found infrared absorptions which they could not at first identify, but which were later found to be due to carbon dioxide.

This was not only unexpected, but also rather disconcerting. Significant amounts of carbon dioxide (CO_2) in an atmosphere will produce a marked greenhouse effect, as the gas does not allow the re-radiated radiation to escape back into space. A dense CO_2-rich atmosphere would inevitably result in a very high surface temperature. The new methods used by Adams and Dunham gave evidence of a state of affairs very different from that pictured 35 years earlier by Johnstone Storey, who had concluded that the Venusian atmosphere was moisture-laden and an almost complete copy of the Earth's!

There were various estimates of the depth of the atmosphere. As recently as 1952 Sir Harold Spencer Jones, the then Astronomer Royal, wrote that the atmosphere "was likely to be somewhat less extensive that that of the Earth", while Henry McEwen, Director of the Mercury and Venus Section of the

Photographs of Venus by W. H. Wright.
(Left) In ultraviolet; (right) in infrared.

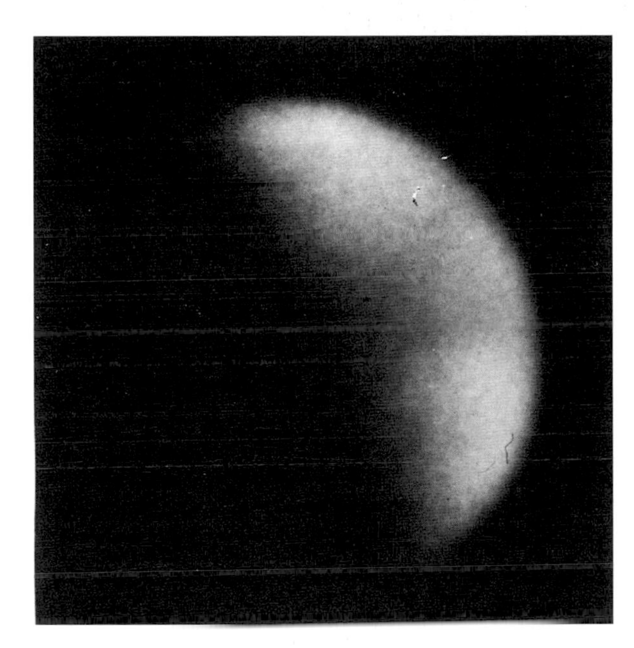

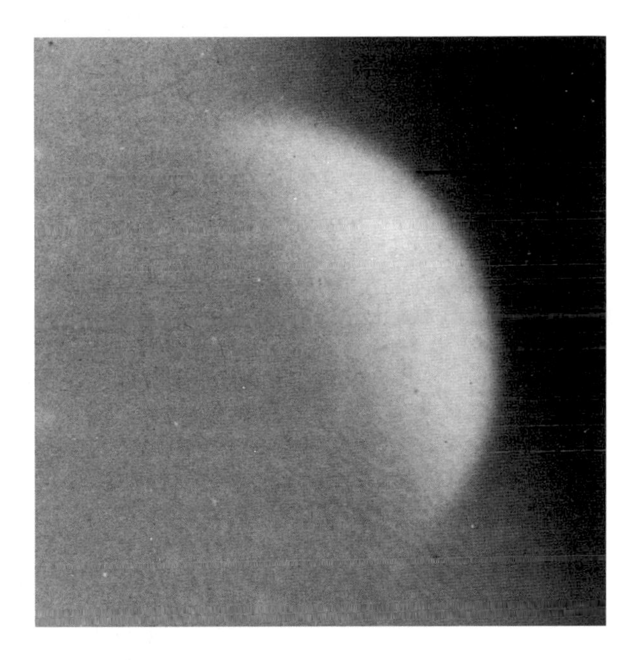

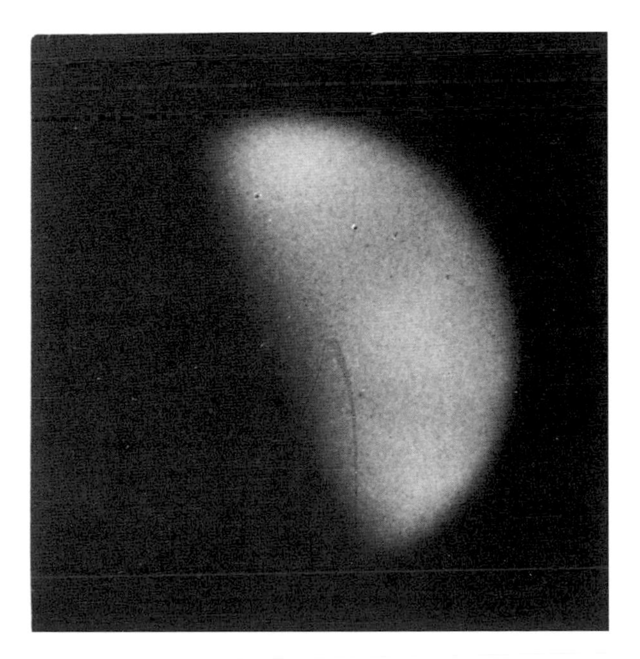

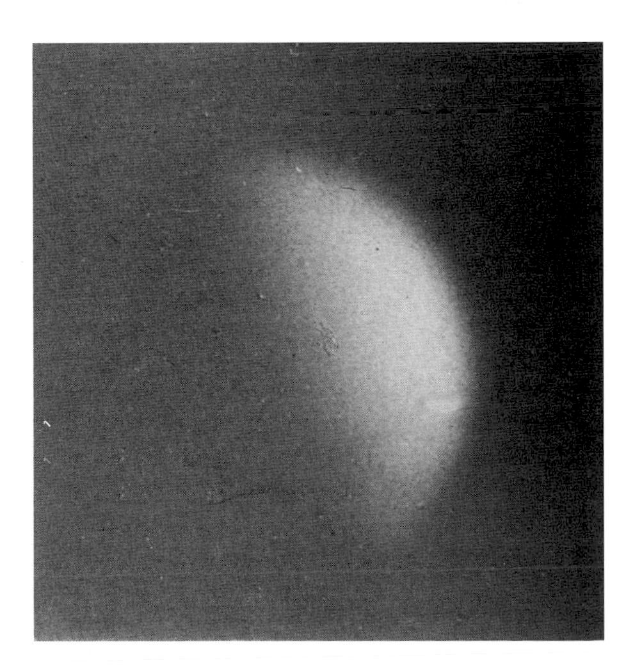

The 100-inch (254-cm) reflector at Mount Wilson.

British Astronomical Association, went to the other extreme and gave an estimated height of 1600 km. In 1937 Rupert Wildt proposed that the clouds might be made up of formaldehyde, a combination of carbon, hydrogen and oxygen (CH_2O) formed under the influence of ultraviolet light emitted by the Sun, but searches for formaldehyde bands in the ultraviolet spectrum of Venus proved to be fruitless. About the same time H. Suess proposed that the clouds contained salts such as sodium chloride and magnesium chloride, produced by the drying-up of former oceans, but there was no supporting evidence.

Real advances were delayed until the 1960s, when data was obtained first by radar and then from spacecraft. In view of this, it may be interesting to look back at some of the theories which were still current – or, at least, seriously considered – earlier on.

The idea of life on Venus was slow to die, and in 1915 C. E. Housden, a strong supporter of Lowell's Martian canal theory, wrote a book about it. His views were, to put it mildly, a little extreme. He held that the 225-day rotation period was valid, so that convection currents would set up between the day and night hemispheres; deposits of ice and snow would be formed just inside the dark hemisphere, and glaciers would drift back into the sunlight, enabling local inhabitants to pump the water back along conduits – these conduits being, of course, the spokes shown on Lowell's maps. It is significant that though the august periodical *Nature*

Artist's impression of Arrhenius' Venus.
[Courtesy of Natural History Museum, London.]

carried a review of the book, Housden's theories were merely criticized rather than ridiculed.

Next in the field was Svante Arrhenius, a Swedish physicist and chemist whose work was good enough to earn him the Nobel Prize in Chemistry in 1903. In a book published in 1918 he gave a vivid and attractive picture of Venus, which he pictured as a world rather in the state of the Earth more than 200 million years ago – in the Carboniferous Period, when coal forests were being laid down and the most advanced life-forms were amphibians; even the great dinosaurs lay in the future. He wrote:

The average temperature there is calculated to be about 47°C ... the humidity is probably about six times the average of that of the Earth, or three times that of the Congo, where the average

A. Oil beneath surface

B. Oil rises to surface to form lakes

C. Oil lakes evaporate into atmosphere

D. Atmosphere laden with evaporate

Diagram illustrating the processes envisaged by Fred Hoyle. Hydrocarbons accumulated beneath the surface (A) eventually rise towards the surface (B); surface lakes of oil then evaporate into the hot atmosphere (C), the latter becoming saturated with the hydrocarbon (D)

temperature is 26°C. The atmosphere of Venus holds about as much water vapour 5 km above *the surface as does that of the Earth at the surface. We must therefore conclude that everything on Venus is dripping wet. The rainstorms, on the other hand, do not necessarily bring greater precipitation than with us. The cloud-formation is enormous, and dense rain-clouds travel as high as 10 kilometres. The heat from the Sun does not attack the ground, but [rather attacks] the dense clouds, causing a powerful external circulation of air which carries water vapour to higher strata, where it condenses into new clouds. Thus, an effective barrier is formed against horizontal air-currents in the great expanses below. At the surface of Venus, therefore, there exists a complete absence of wind both vertically, as the Sun's radiation is absorbed by the ever-present clouds above, and horizontally, due to friction. Disintegration takes place with enormous rapidity, probably about eight times as fast as on Earth, and violent rains carry the products speedily downhill, where they fill in the valleys and the oceans from all river mouths.*

A very great part of the surface of Venus is no doubt covered with swamps, corresponding to those on the Earth in which the coal deposits were formed, except that they are about 30°C warmer. No dust is lifted high into the air to lend it a distinct colour; only the dazzling white reflex from the clouds reaches the outside space and gives the planet its remarkable, brilliantly white lustre. The powerful air-currents in the highest strata of the atmosphere equalize the temperature differences between poles and equator almost completely, so that a uniform climate exists all over the planet analagous to conditions on the Earth during its hottest periods.

The temperature on Venus is not so high as to prevent a luxurious vegetation. The constantly uniform climatic conditions which exist everywhere result in an entire absence of adaptation to changing exterior conditions. Only low forms of life are therefore represented, mostly no doubt belonging to the vegetable kingdom; and the organisms are of nearly the same kind all over the planet. The vegetative processes are greatly accelerated by the high temperature. Therefore, the lifetime of organisms is probably short. Their dead bodies, decaying rapidly, if lying in the open air, will fill it with stifling gases; if embedded in the slime carried down by the rivers, they speedily turn into small lumps of coal, which later, under the pressure of new layers combined with high temperature, become particles of graphite.

... The temperature at the poles of Venus is probably somewhat lower, perhaps by about 10°C, than the average temperature on the planet. The organisms there should have developed into higher forms than elsewhere, and progress and culture, if we may so express it, will gradually spread from the poles toward the equator. Later, the temperature will sink, the dense clouds and the gloom disperse, and some time, perhaps not before life on Earth has reverted to its simpler forms or has even become extinct, a flora and fauna will appear, similar in kind to those which now delight our human eye, and Venus will then indeed be the "Heavenly Queen" of Babylonian fame, not because of her radiant lustre alone, but as the dwelling-place of the highest beings in our Solar System.'

Arrhenius found some supporters, though in general astronomers tended to be somewhat sceptical about a Carboniferous Venus.

1955 turned out to be a vintage year for Venus theorists, and three different ideas were put forward. First came Sir Fred Hoyle, who made the startling suggestion that Venus might well have oceans – not of water, but of oil! His reasoning was as follows:

'Suppose an enormous quantity of oil were to gush to the Earth's surface; what would the effect be? The oil, consisting as it does of hydrocarbons, would proceed to absorb oxygen from the air. If the amount of oil were great enough all the oxygen would be removed. When this happened the water vapour in our atmosphere would no longer be protected from the disruptive effect of ultraviolet light from the Sun. So water vapour would begin to be dissociated into separate atoms of oxygen and hydrogen. The oxygen would combine with more oil, while hydrogen atoms would proceed to escape altogether from the Earth into space. More and more of the water would be dissociated and more and more of the oil would become oxidized. The process would only come to an end when either the water or the oil became exhausted. On the Earth it has been clear that water has been dominant over oil. On Venus the situation seems to have been the other way round; the water has become exhausted and presumably the excess of oil remains – just as the excess of water remains on Earth.

This possibility has an interesting consequence. In writing previously about these clouds I said that the only suggestion that seemed to fit the observations was that the clouds are made up of fine dust particles. To this suggestion we must now add the possibility that the clouds might consist of drops of oil – that Venus may be draped in a kind of perpetual smog ...'

And with regard to the slow rotation:

'It is thus reasonable to suppose that the slowing-down of Venus can be explained by the friction of the tides – if Venus possesses oceans, but not I think otherwise. Previously the difficulty was to understand what liquid the oceans were made of. Now we see the oceans may well be oceans of oil. Venus is probably endowed beyond the dreams of the richest Texas oil-king.'

In view of the present world situation, this would make Venus an attractive target! But at the same time two eminent American astronomers, Fred Whipple and Donald Menzel, produced a totally different picture. This time the Venusian oceans were nothing more nor less than ordinary water, with clouds made up of H_2O. The theory was based mainly on a series of measurements of the polarization of the light of Venus made earlier by Bernard Lyot, at Meudon Observatory. Lyot found that only water droplets agreed reasonably well with the variation in polarization with scattering angle on Venus, so that Whipple and Menzel concluded:

'Lyot's polarization measures indicate that water vapour droplets fit the data satisfactorily. The droplets of the Venusian clouds are uniform in dimension and large for high-level airborne dust. No one has been able to suggest, in place of water droplets, a likely substitute material that would be both available and agree with the polarization and other reflection characteristics observed on the clouds.'

They therefore maintained that the surface of Venus was likely to be completely covered with water.

On this theory, a dense atmosphere consisting largely of carbon dioxide could not exist upon an Earth-like planet with continents occupying a large fraction of the surface area; the carbon dioxide would be fixed in the rocks to form carbonates, because of the chemical reaction with silicates in the presence of water. If land-masses were virtually absent, however, the fixation of carbon dioxide could not continue after the formation of a thin buffer layer of carbonates and this was the reason for supposing that the surface must be essentially oceanic. There was one rather bizarre corollary. Presumably the atmospheric CO_2 would have fouled the oceans, giving Venus seas of nothing more nor less than soda-water.

Of course, the idea of Venusian oceans was not new. Another Nobel laureate, Harold Urey, had commented that "the presence of carbon dioxide in the planet's atmosphere would be very difficult to understand unless water were originally present, and it would be impossible to understand if water were present now, so that the former oceans dried up." Urey also maintained that mountain-building was likely to be on a very reduced scale; "Venus may have produced its mountains in the past and they may have been eroded to sea-level perhaps in the distant past, and it may be that the Earth has sufficient size to continue this activity at the present time." Urey also rejected the idea of volcanoes on Mars – though we now know that one Martian volcano, Olympus Mons, towers to three times the height of Mount Everest!

In very early times the Earth's atmosphere contained much more carbon dioxide and much less free oxygen that it does now, so that if we could enter a time machine and send ourselves back to the primitive world we would promptly suffocate. Free oxygen became abundant only when plants spread on to the land areas, and the process of photosynthesis became all-important. If Venus were a watery world, why should not life have begun there, just as it did in the warm seas of the ancient Earth? The prospect was intriguing, and when Whipple and Menzel proposed their marine theory, it certainly could not be rejected out of hand.

But could we actually glimpse the true surface through the clouds? Urey believed so, and even believed that there might be shallow seas; the dark areas seen on the planet could well be low, swampy land covered with vegetation. Next came G. A. Tikhov, of the USSR, who founded what he called the study of "astrobotany"; his Venus was not in the Carboniferous period, but more nearly resembled the Earth of a hundred million years ago. He wrote:

'Now we can already say a few things about the vegetation of Venus. Owing to the high temperature of the planet, the plants must reflect all the heat rays, of which those visible to the eye are the rays from red to green inclusive. This gives the plants a yellow hue. In addition, the plants must radiate the red rays. With the yellow, this gives them an orange colour... Our astrobotanical conclusions concerning the colour of vegetation on Venus find certain confirmation in the observations of academician N. P. Barabashov. He found that in those parts of Venus where the Sun's rays possibly penetrate the clouds to be reflected by the planet's surface, there is a surplus of red and yellow rays. Barabashov believes that the surface of Venus is to a certain degree specular, and that the yellow and red rays pass more easily through the clouds than do the rays at the blue end of the spectrum. I in my turn wish to add that here a certain part may be played by the

Photograph of Venus with Earth-based telescopes, by H. E. Ball.

Cut-away diagram of probable circulation of Venus's atmosphere.

vegetation of Venus. Thus we get the following gamut of colours: on Mars where the climate is rigorous the plants are of blue shades. On Earth where the climate is intermediate the plants are green, and on Venus where the climate is hot, the plants have an orange colour.'

Venus hid its secrets well, and it began to look as though we could learn little more until it became possible to dispatch space-craft. Ironically, the most informative work, carried out by the French astronomers Boyer and Guèrin in 1961, caused very little interest at the time, and was even regarded with considerable mistrust. Boyer and Guèrin, at the Pic du Midi, carried out a long photographic study of Venus, and recorded a characteristic Y-shaped feature centred on what they (correctly) believed to be the equator. It persisted throughout the observations, and from its movements it was deduced that the rotation period of the upper clouds could be no more than 4 days, in a retrograde sense. This seemed improbable, as all other work indicated that the rotation period was long. Yet in the event it proved to be accurate; the uppermost clouds really do race round at this speed.

At last, in the 1960s, came the first space missions. What would they find – a raging dust-bowl, an ocean teeming with life, or something quite different? Astronomers and geologists were very anxious to find out.

MISSIONS TO VENUS 4

The first probe to approach Venus was the Russian craft, Venera 1, with which contact was lost early in 1961. It was not until 1962 that Mariner 2 flew past the planet, sending back much vital information. Subsequently, the Russian Venera program has not only landed probes on Venus but imaged its surface by radar and made atmospheric measurements. The American Pioneer-Venus probes provided the first global maps of the planet.

The Space Age began on 4 October 1957, as Russia's Sputnik 1 soared aloft, and entered a closed orbit round the Earth. It was football-sized, and carried little apart from a radio transmitter which sent back the never-to-be-forgotten "Bleep! bleep!" signals which caused such interest all over the world – plus surprise and, at least in the USA, a considerable amount of alarm. The Americans had already announced plans to launch an artificial satellite, but political manoeuvring had delayed it, and only after Sputnik 1 did they authorize one of their leading scientists, Wernher von Braun, to send up a satellite (which, a mere four months later, he duly did).

At this stage – the start of the now mercifully forgotten "Space Race" – the Russians

Venera 1. This was the first probe to approach the hostile planet.

had a clear lead. Sputnik 1 was followed by heavier satellites, and then the first automatic probes to the Moon, one of which, Luna 3, made a circum-lunar trip in October 1959 and sent back the first pictures of the far side. Next, in 1961, Yuri Gagarin became the first of all astronauts. It was only natural that the Soviet planners should start to consider true interplanetary vehicles, and Venus, our nearest neighbour in space after the Moon, was an obvious target – particularly as at that stage it was still believed to be a reasonably friendly world, and as a potential colony rather more promising than Mars.

Sending a probe to another planet is not nearly so straightforward as might be thought. What cannot be done is to wait until the Earth and the target planet are at their closest, and then simply fire a rocket across the gap. This would mean using propellant for the entire journey, and no vehicle could possibly carry enough; almost the whole trip has to be done unpowered, in a transfer orbit, using the Sun's gravity. The probe is sent up via a powerful launch vehicle, a step rocket, and then either speeded up or slowed down relative to the Earth, depending upon where it is meant to go. With Venus, the motors on the probe itself have to be used to slow down the orbital speed. The probe will then start to swing in towards the Sun, and calculations have to ensure that it arrives at the orbit of Venus at the right moment for a rendezvous.

The main weakness in the Russian programme in the early days was long-range communication, and their first foray to Venus, with Venera 1, emphasized this. From all accounts the launch, from the cosmodrome at Baikonur, was faultless, and Venera 1 went on its way. Unfortunately contact was lost when the probe was only 7,500,000 km from Earth, and was never regained. Venera 1 may have by-passed Venus at around 100,000 km in May 1961, but its final fate will never be known; no doubt it is still orbiting the Sun, but we have no hope of finding it again. Presumably it was not intended to land, and was simply a fly-by, but in any case it was more than three years before the Russians made another attempt, and by then the Americans had achieved a major triumph with Mariner 2. A full list of Venus missions is given in Appendix 2.

America's first Venus mission, Mariner 1, was dispatched on 22 July 1962. It was a prompt and total failure, plunging into the sea a few minutes after take-off. (Apparently someone had forgotten to feed a minus sign into a computer – a slight mistake which cost $10 million.) It was not an auspicious beginning, but Mariner 2, launched on the following 27 August, more than made up for its predecessor's ignominious end.

Again the probe was to travel in a transfer orbit, making use of solar gravity and "coasting" for most of its journey. Mariner went up from Cape Canaveral; as usual at this time, the bottom stage of the launch vehicle was an Atlas rocket. The launcher ascended almost vertically, and then headed off in the general direction of South Africa. The second stage, an Agena rocket, then took over, and the Agena–Mariner combination entered what is termed a parking orbit, at a mean altitude of 185 km above the surface of the Earth and moving at a velocity of 29,000 km h^{-1}. As it reached the African coast, the Agena fired again, and the total velocity reached 41,034 km h^{-1}, which is more than the escape velocity at that altitude.

Remember that for a Venus mission, the direction of escape has to be opposite to that in which the Earth is moving round the Sun. It was so on this occasion. As Mariner 2 moved away, it was slowed down by the Earth's gravity until, when it was 965,000 km away, its velocity relative to Earth had dropped to 11,090 km h^{-1}; in other words, Mariner 2 was moving round the Sun at a rate 11,090 km h^{-1} less than the Earth. It therefore started to swing inward toward the Sun, gathering speed as it went. Meanwhile, the Agena rocket had been separated from the Mariner probe, rotated through a wide angle and fired for the last time, so that it was put into an entirely different orbit. Nobody knows what happened to it, and nobody cares; it had completed its task, and on its journey to Venus it would simply have been a nuisance. Henceforth Mariner 2 was on its own.

It would have been over-optimistic to hope for complete accuracy at what was to all intents and purposes a first attempt (Mariner 1 can hardly be counted, as its demise was so quick). A mid-course correction was made on 4 September, but even so the minimum distance from Venus was 34,833 km instead of the planned 16,000 km. Fortunately the instruments had been designed to operate over a considerable range, so that the error was not ruinous.

The date of closest approach was 14 December. All went well; contact was maintained without difficulty, and a tremendous amount of information was returned, though

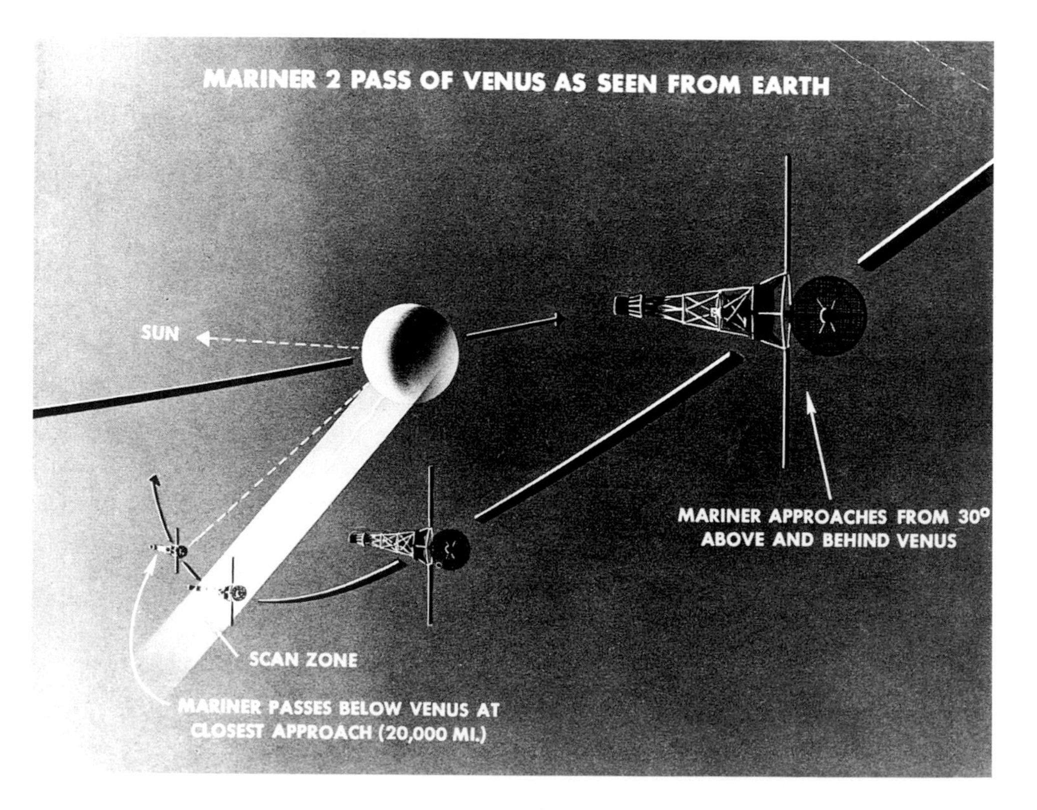

The orbit and approach of Mariner 2 to Venus. By 1963, the data sent back by this small craft had confirmed Venus to be an extremely hot world with a choking atmosphere of CO_2 and no magnetic field.

Mariner carried no camera. It took some time to analyse and interpret the results, but by February 1963 the situation had become fairly clear, and it was obvious that many of our cherished ideas about the planet would have to be jettisoned.

First, and most important of all, there was the question of the temperature. Instead of being merely tropical, Venus proved to have a surface so hot that the Whipple–Menzel marine theory had to be ruled out at once; liquid water could not possibly exist in such heat, even under a dense atmosphere. There was no detectable magnetic field; the atmosphere was made up chiefly of carbon dioxide, as had been expected; and a very long rotation period, already indicated by Earth-based radar, was confirmed. For the first time it was demonstrated that the rotation period is longer than the Venusian year, and that the rotation is in a sense opposite to that of the Earth or Mars, so that if an observer could stand on the planet and the see the Sun it would rise in the west and set in the east.

In passing, the reason for Venus's retrograde spin is still not known. The popular view is that at an early stage the planet was struck by a massive body and "tipped over", so to speak. While this may seem implausible, such bodies may have been fairly common during the early days of Solar System history (another may have been responsible for stripping away much of Mercury's mass.)

Contact with Mariner 2 was finally lost on 4 January 1963, at a distance of 87 million kilometres. It had achieved everything that its makers had hoped, even if the results had been rather depressing to those astronomers who had been hoping to find a moderately friendly environment.

The next attempt was Zond 1, a Russian probe launched in April 1964. Not much is known about this; at this time the Soviets were decidedly secretive, and all that can really be said about it is that it failed. Next came Venera 2, in late 1965, which flew by Venus at 24,000 km but apparently sent back few data; contact was lost before the critical period, and Venera 2 continued in solar orbit unseen and unheard. But Venera 3, launched four days later, was very different. According to statements from Moscow, it actually landed on the planet on 1 March 1966.

The landing was not controlled, so that Venera 3 must have destroyed itself on impact. Contact was lost just before the descent through the atmosphere, and no scientific results were obtained; on balance, therefore, Venera 3 must be classed as another failure. At the time it attracted some criticism, inasmuch as there seemed to be a real risk that experiments of this sort might carry Earth contamination to Venus. Unfortunately it is impossible to ensure that every probe is completely sterilized, and any Earth bacteria might well ruin the chance of examining another world in its mint condition, so to speak. Obviously all precautions are taken, and the danger is slight, but it is not nil.

The next development came in 1967. The Soviet team sent up Venera 4 on 12 June, and America's Mariner 5 followed two days later. The aims of the two vehicles were not the same. Venera 5 was an intended soft-lander, while Mariner 5 was a fly-by probe due to by-pass Venus and continue in solar orbit.*

*In case you are wondering about Mariners 3 and 4, both were Mars probes; No. 3 failed, but No. 4 succeeded, and in July 1965 sent back the first close-range pictures from the cratered uplands of Mars. Altogether it returned 21 images, plus much miscellaneous information, and contact with it was not lost until 20 December 1967. Its closest approach to Mars was 9789 kilometres.

The pattern of descent and landing for the Venera 9 and 10 probes.

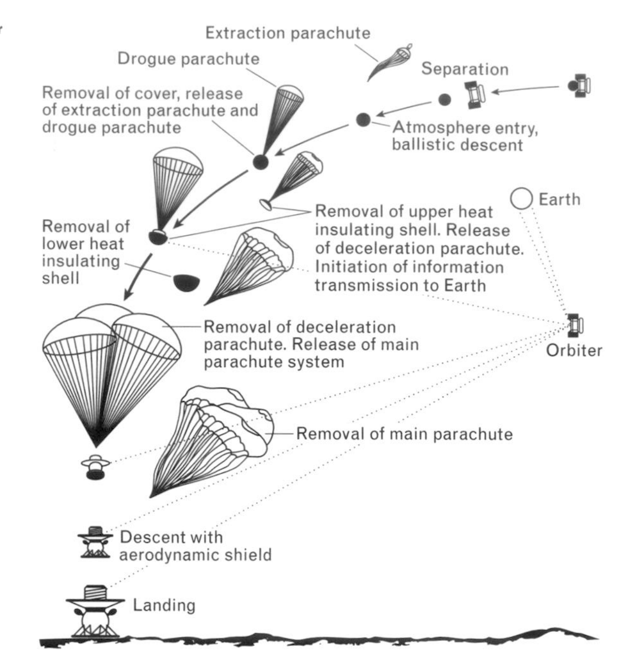

Real success came with Venera 7, which landed on the planet on 15 December 1970. It had been modified to withstand the most extreme surface temperatures and pressures which had been estimated from the previous missions, and it should have been able to tolerate a pressure of 180 atmospheres and a temperature of 800°C. It seems to have come down in the region of Navka Planitia, and signals were received for over 20 minutes after impact.

The difficulties were very great. The intense heat meant that the probe had to be chilled as much as possible before plunging into the dense atmosphere, and of course everything had to be automatic; once the signal to start the entry had been given, the operators on Earth had absolutely no control over what was happening. But the results were starting to look encouraging; Venera 8, in 1972, repeated the success of its predecessor and transmitted for almost an hour after landing. The stage was set for an attempt to send back pictures from the solid surface. First, however, a word or two about America's next foray.

Mariner 10, of 1973, was essentially a Mercury probe; up to that time no spacecraft had been sent to that strange little planet, and not much was known about its surface details, though it was known to have virtually no atmosphere, and it seemed likely that its surface would not be too unlike the Moon. Mariner 10 was dispatched from Cape Canaveral on 3 November 1973, and encountered Venus at a distance of 5800 km, on the following 4 February. In fact Venus was merely incidental on this occasion; it was used simply to alter Mariner's orbit and send it on to rendezvous with Mercury – a procedure known as a gravity-assist. However, the chance to make further studies of the upper atmosphere of Venus was too good to be missed, and Mariner 10 sent back good images over a period of 8 days, showing detailed pictures of the upper clouds and fully confirming the 4-day rotation of the atmosphere. It also established that there is very little difference in temperature between the day and night hemispheres.

Mariner 10 subsequently acquitted itself well. It made three active passes of Mercury – two in 1974 and one in 1975 – and sent back excellent pictures as well as a great deal of miscellaneous information. Contact was finally lost on 24 March 1975. No doubt it is still circling the Sun, and still making regular close passes of Mercury, though so far as we are concerned it is dead.

It had long been the Russian aim to obtain a picture direct from the Venusian surface, and they succeeded with Veneras 9 and 10, both of which touched down in October 1975 and sent back one image each. It was a spectacular triumph – and rather surprising inasmuch as the Soviet planners had had no

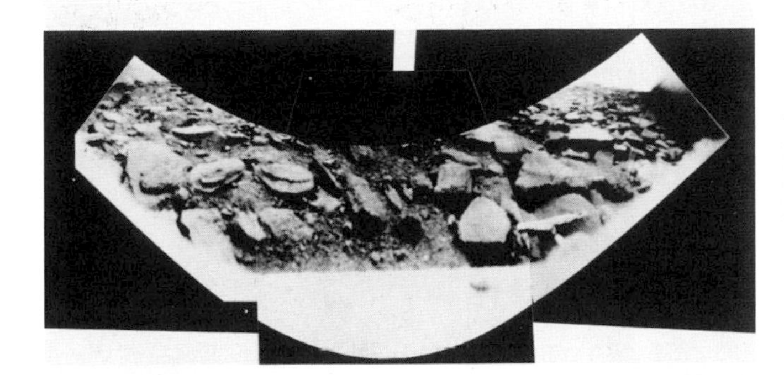

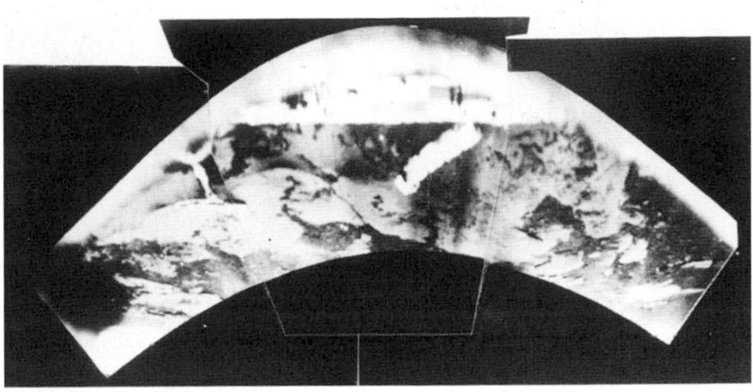

The surface of Venus from Venera 9 (left) **and Venera 10** (right). **The images have been transformed to correct for distortion.** [Courtesy of James Head, Brown University, Rhode Island.]

success at all with Mars, which logically should have been a much easier target.

Each Venera consisted of two parts: an orbiter, and a lander. The two sections crossed space together, and were separated while in a closed orbit round the planet; the landers were brought down through the atmosphere and, after aerodynamic braking, fell slowly by parachute from 65 to 50 kilometres, giving prolonged coverage of the cloud layers. At 50 kilometres the parachutes were released, allowing for more rapid descent through the intensely hot lower atmosphere to be controlled by a braking shield. The velocity of impact was no more than 7 metres per second. Obviously it was hopeless to expect transmissions to go on for long; in the event Venera 9 sent back signals for 56 minutes after arrival, and Venera 10 for ten minutes longer.

The view from Venera 9 was remarkable by any standards. The first shock was the amount of light available. It had been expected that the atmospheric clouds would turn Venus into a very gloomy world indeed, but actually the Russians commented that the level of illumination was about the same as that in Moscow at noon on a cloudy winter day. Though the probes had taken floodlights with them, no artificial lighting had to be used. Neither was the atmosphere "super-refractive". Suggestions had been made that the great density of the lower atmosphere would bend light-rays very sharply, and one bizarre comment was that an astronaut who could step outside his spacecraft and look ahead would see the back of his neck! Of course this was a deliberate exaggeration; but certainly there were no effects of this kind, and features up to a hundred metres away from the ground vehicle were clearly shown.

The next surprise was the lack of dust-sized material; apparently neither lander kicked up any sort of cloud. Venera 9 showed a rock-strewn landscape; the rocks were sharp-edged, and most were a metre or so across. One Soviet scientist (Mikhail Marov) described the scene as a "stony desert", and there was no evidence of violent erosion. This was unexpected, because even a slow wind would have tremendous force in so dense an atmosphere. Measurements showed that the windspeed was of the order of seven knots at most, though the Russians pointed out that this would be as devastating as the effect of a sea-wave upon a terrestrial cliff during a 40-knot gale.

The scene from Venera 10 was much the same, but showed large slabby rock outcrops, with less scattered fragments than at the Venera 9 site. Moreover the rocks were distinctly smoother, and the general outlook was of a landscape appreciably older.

The quality of the pictures was remarkable, considering the conditions under which they had been taken. To obtain transmissions for an hour or so from each probe was also a major achievement. The landing manoeuvres themselves had been difficult, and each took over an hour from the moment of entry into the upper cloud-layer. It is now known that the cloud-base ends

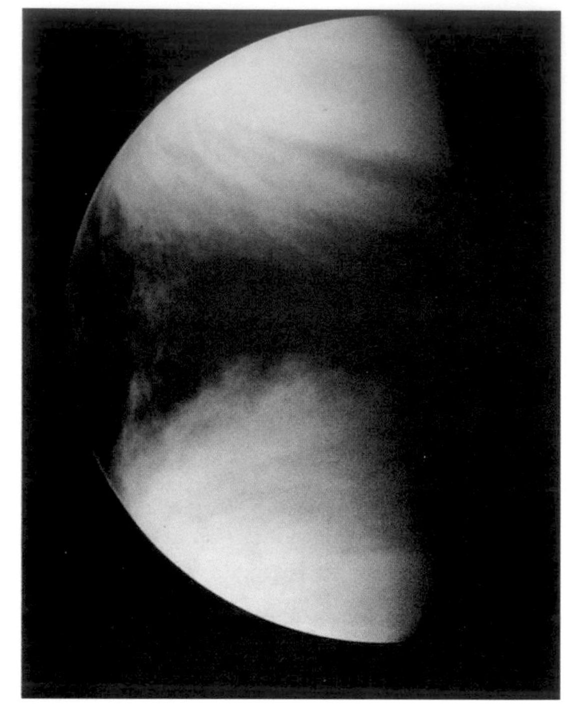

1. 30 December 1978

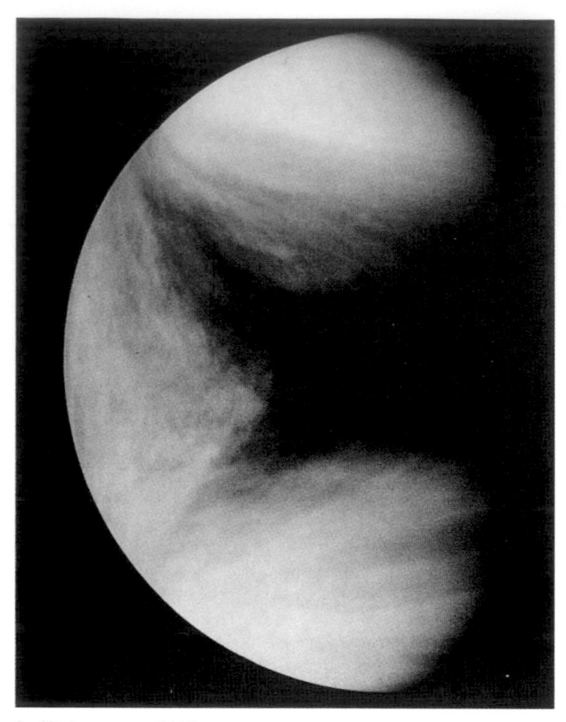

2. 10 January 1979

Four images of Venus from Pioneer-Venus in 1978–9. The Y-shaped cloud pattern is believed to be due to eddies in the Venusian atmosphere.

well above the actual surface, and below lies what may be termed a superheated smog.

There was also corrosion to be taken into account. It had already been established that the clouds contain large amounts of sulphuric acid, and all in all conditions on Venus could hardly have been worse; indeed, they were remarkably like the conventional idea of hell.

Veneras 11 and 12 were essentially similar to their predecessors, with both orbiter and lander systems, but in each case the television cameras failed, and no images were received. However, Veneras 13 and 14 more than made up for this disappointment. Colour images were received from the surface, and soil analyses were attempted for the first time. Venera 13 came down in the region of Phoebe Regio, and its lander transmitted for a record 127 minutes after arrival; the temperature was found to be 457°C, and the pressure 94 atmospheres. Analyses showed that in the surface materials, highly alkaline potassic salts were much in evidence, though the general composition of the rocks was basaltic.

Meanwhile, the American Pioneer-Venus mission had begun its operations. It involved two spacecraft; an orbiter, and a multiple-entry probe ("multiprobe"), which consisted of one large and three small entry probes mounted on a "bus", which was rather similar in design to the orbiter. The multiprobe concept had the advantage of being able to carry out simultaneous studies of the atmosphere in various locations, on both the day-side and night-side. The orbiter reached Venus on 4 December 1978. Its distance from Venus ranged between 150 km and as much as 66,900 km; the period of revolution was just over 24 hours, and the inclination to the Venusian equator was between 105° and 106°.

The five separated spacecraft of the multiprobe mission (bus, large sounder probe, small sounder probe, small day probe and small night probe) landed on 9 December. Their entry into the atmosphere, at about 200 km above the surface, was staggered over a period of ten minutes to make ground reception easier. After entry, atmospheric braking began well above 100 km, reaching

3. 14 January 1979

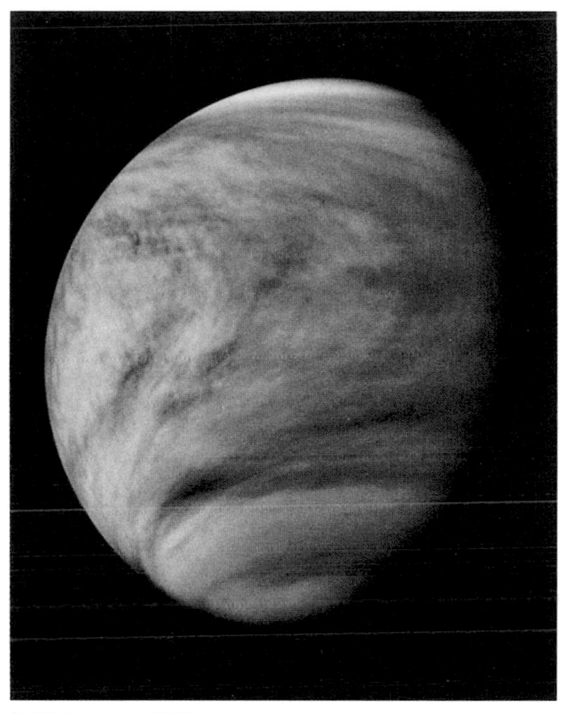

4. 18 January 1979

peak deceleration of 500 to 600 g near 78 km. At 68 km the main parachute of the large probe opened, and the probe descended slowly through the cloud layers. After 17 minutes, at around 47 km, the parachute was jettisoned, allowing the probe to fall aerodynamically until it hit the surface 39 minutes later.

The small probes did not carry parachutes, and fell freely from 60 km above the surface. They were not designed to operate after impact, though in fact one of them – the day probe – did so, and continued to send back signals for 67 minutes after arrival.

The main tasks of all these entry probes was to study the composition and structure of the atmosphere, and on the whole the investigations were successful, though several instruments failed for reasons which have never been properly understood. The "bus" had no braking mechanism, and simply burned away in the upper atmosphere, but the orbiter continued to send back data for a long time, and did not finally "die" until

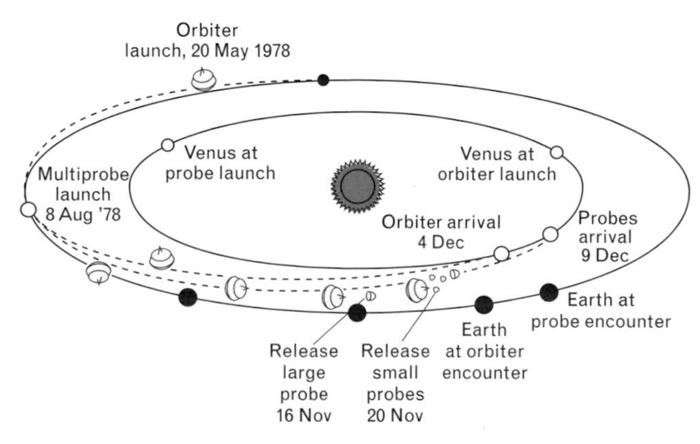

The flight path of the Pioneer-Venus spacecraft. This also shows the release points of the large and small descent probes.

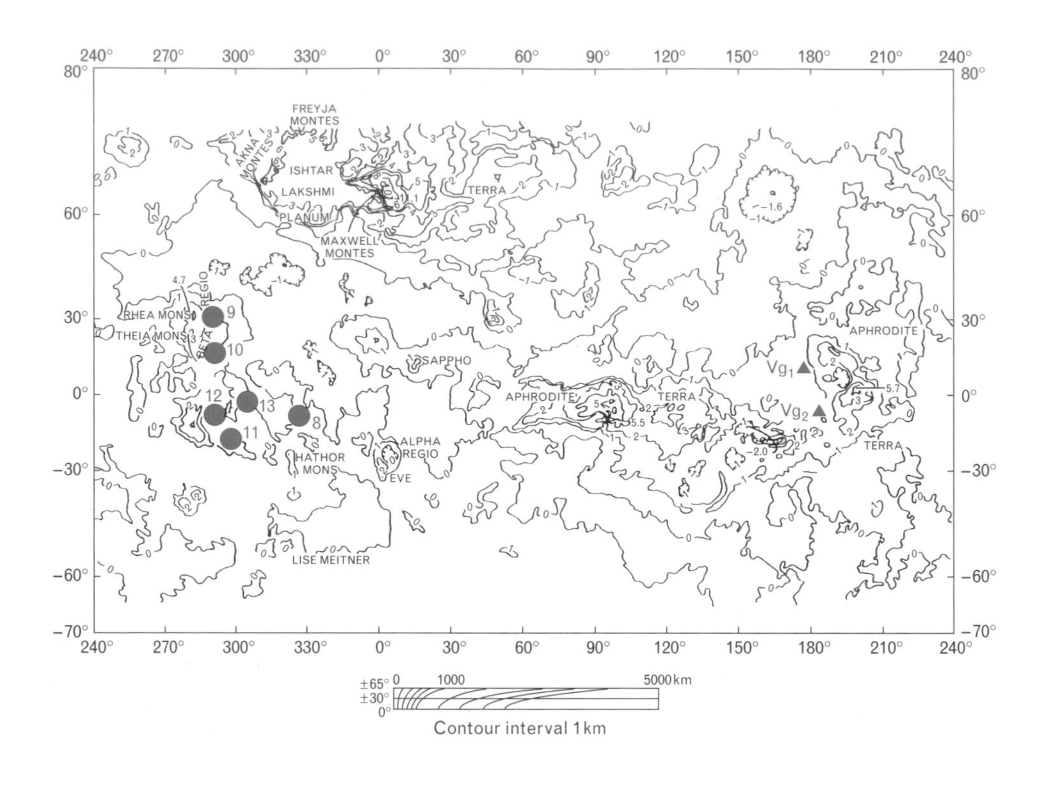

Map of the landing sites for successful Venus probes. Venera landing sites represented by 8–13; Vega landers located at Vg1 and Vg2.

9 October 1992. The latter was essentially a radar mapping spacecraft, and provided the first really detailed physiographic map of the surface of the planet, together with gravity data. The latter was obtained by accurately tracking the spacecraft's motion around the planet, the craft speeding up when over relatively dense regions and slowing down above less dense areas.

The latest Russian mission, to date, was a decidedly unexpected one. In 1986 Halley's Comet came back to the Sun for the first time in 76 years, and created a tremendous amount of interest, not only among the astronomical community but also in the general public. Though it was badly placed for observation, and never became brilliant – as it had been in 1835 and 1910 – this was the most important return on record, since spacecraft could be used. The Americans cancelled their mission on the grounds of cost, but a whole armada was sent to the comet; two Russian probes, two Japanese, and one European. The European vehicle, Giotto, built in the UK, penetrated the

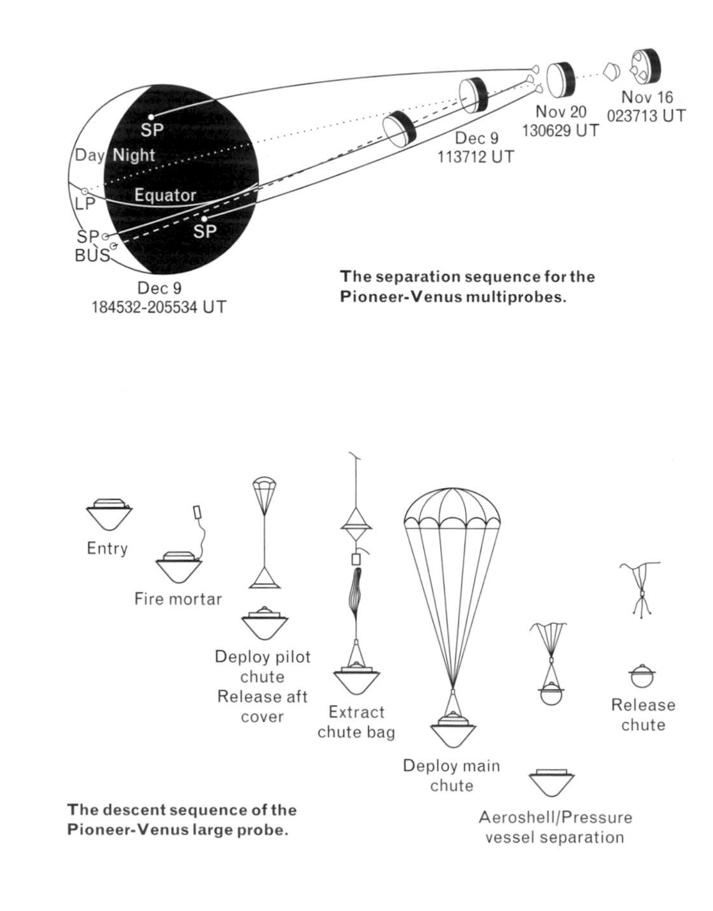

The separation sequence for the Pioneer-Venus multiprobes.

The descent sequence of the Pioneer-Venus large probe.

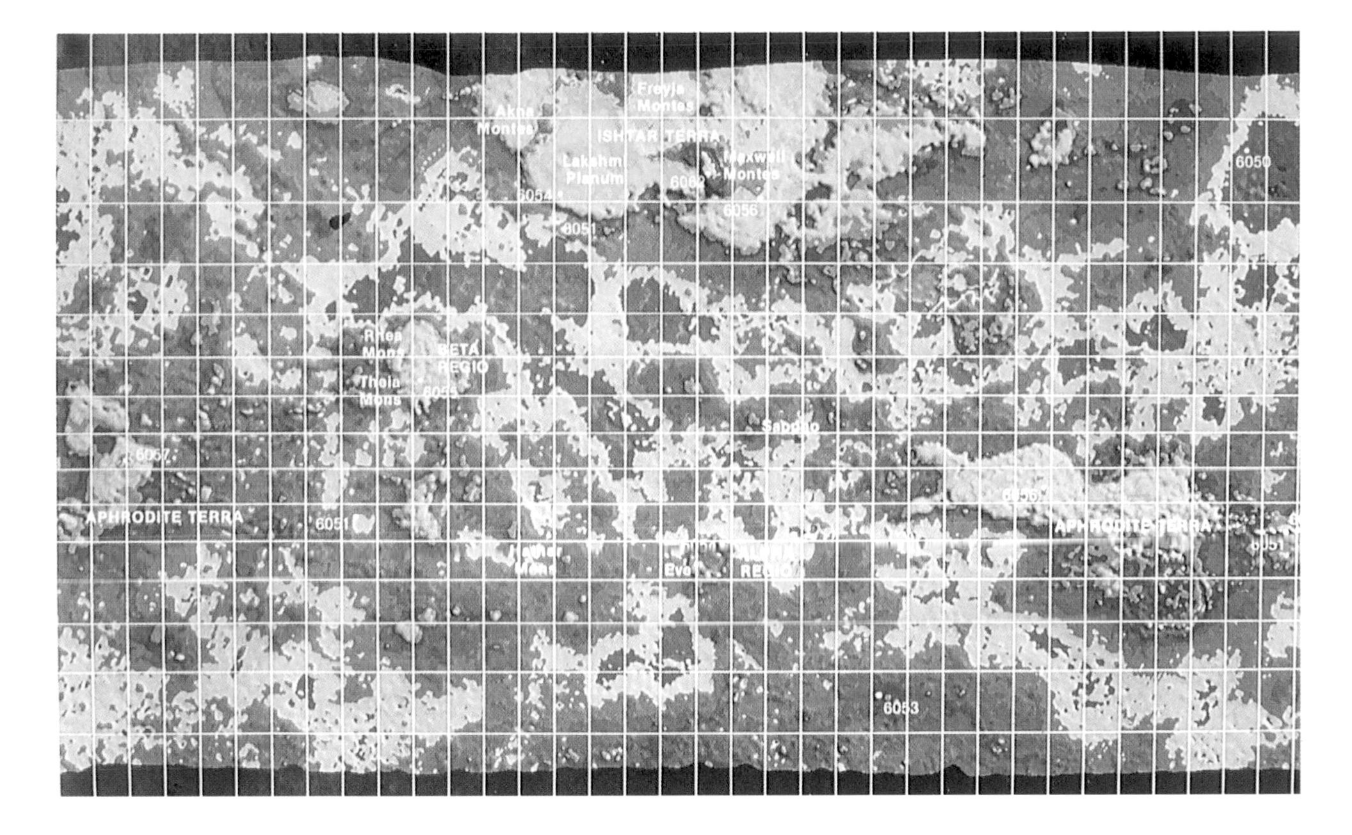

Pioneer-Venus colour-coded physiographic map of Venus.

comet's head and sent back close-range pictures of the cratered icy nucleus; to many people's surprise it survived the encounter, and went on to rendezvous with a smaller comet, Grigg-Skjellerup, in July 1992.

More or less at the last moment, the Russian scientists realized that their twin probes, Vega 1 and Vega 2, could pass by Venus en route for the comet, and it seemed only sensible to incorporate extra equipment to take advantage of the opportunity which Nature had provided. The main experiments involved dropping balloons into Venus's atmosphere. The balloon from Vega 1 entered the atmosphere at a speed of 11 km s^{-1} over the night-side on 10 June 1985, and was tracked for 56 hours as it drifted on to the day-side; the second balloon was equally successful. It was found that buffetting from the winds was greater than had been expected; for example the Vega 2 balloon encountered a sudden down-draught when it was above the upland area of Aphrodite Terra, on the day-side, and plummetted nearly3 km in half an hour, with lateral gusts of nearly 60 km h^{-1}. It was also found that the Vega 1 balloon registered a temperature distinctly higher than that of its twin, even though the two floated at much the same height and at similar latitudes to either side of the equator – Vega 1 to the north, Vega 2 to the south. Altogether the first balloon was tracked for over 11,000 km, and the second for almost as far. Small landers were also dropped, confirming the very high surface pressures and temperatures and analysing the surface rocks.

By 1989 all our ideas about Venus had been revolutionized. The rotation period had been defined as 243.16 days at the surface, 4 days in the uppermost clouds; the atmosphere consisted of 96.5% carbon dioxide and 3.5% nitrogen, with traces of other gases; the surface pressure was around 90 times that at the Earth's surface; the temperature of the order of 480°C; the magnetic field was negligible, and the clouds were rich in sulphuric acid. Yet our knowledge was still incomplete, and this prompted the dispatch of the most ambitious probe to date: Magellan.

MAPPING VENUS 5

Mapping Venus requires the use of radar, whose long wavelengths can penetrate the dense Venusian atmosphere and reach the surface. Both Earth-based and orbiting instruments have been used successfully in mapping the planet and have revealed wide variations in surface roughness, altitude and landform morphology.

Since the atmosphere of Venus is opaque to radiation at visual wavelengths, scientists have had to develop alternative kinds of imaging systems. Radar provides the ultimate weapon against the Venusian "shield". Radar is an imaging system that operates at wavelengths much longer than those which characterize the visual part of the electromagnetic spectrum and can penetrate clouds with consummate ease. Mapping of the Venusian surface by radar has been undertaken both from orbiting probes and from the Earth.

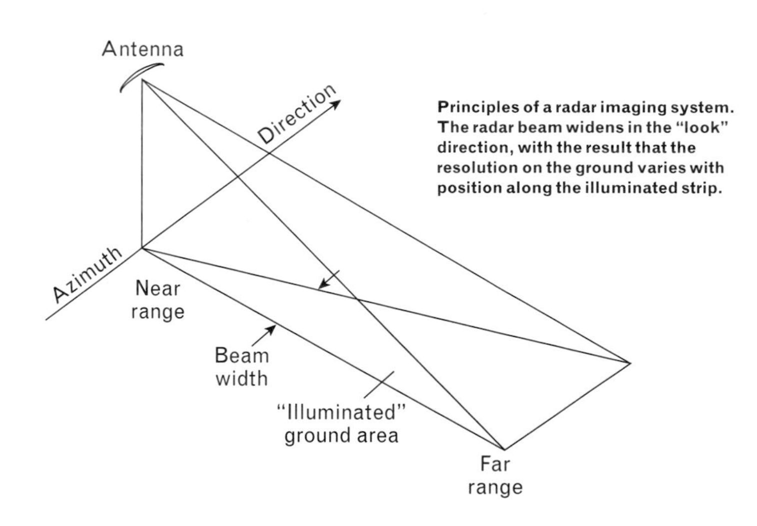

Principles of a radar imaging system. The radar beam widens in the "look" direction, with the result that the resolution on the ground varies with position along the illuminated strip.

Radar and radar imaging systems

Whereas optical telescopes utilize that part of the electromagnetic spectrum we know as the visual band (0.4–0.7 μm) radar imaging devices not only provide their own source of energy but also generate radiation at wavelengths between 0.8 and 70.0 cm, much longer than visual ones. At such long wavelengths – which characterize what is termed the *microwave* region – the radiation is hardly affected either by clouds or by darkness and can penetrate through several metres of solid rocks. It has obvious advantages to the planetary scientist interested in Venus.

Radar is an *active* imaging system, that is, one which produces its own pulse of radiation. It is fundamentally different from *passive* systems which use the Sun's rays for illumination. With radar, a concave antenna focusses the beam into the required form and generates a microwave pulse which "illuminates" the surface over which it passes, the illuminated region being a long and narrow strip normal to the spacecraft's flight path. The "look direction" of the system is

The larger the antennae the more powerful is the system and therefore the greater is the potential for obtaining high-resolution imagery. In order to optimize on size of aperture, radar antennae tend to be fixed, so as not to limit their size. In *real aperture* imaging systems, an antenna is directed towards the surface and collects images of as high resolution as it can within the constraint of its own dimensions or "real" aperture. To collect very high-resolution imagery by this method currently is not possible, because very large antennae cannot be carried onboard spacecraft.

Fortunately, there is a way around this obstacle which utilizes the well-known **Doppler principle**: because the antenna is in motion while it is receiving echoes from a planet's surface, special computer processing can simulate a much larger antenna. Thus, the distance a spacecraft moves while a surface feature is within the radar system's field of view, effectively determines the functional size of the aperture. This appropriately is called *Synthetic Aperture Radar*, or SAR. The technique demands that the radar system can transmit and receive multiple outgoing pulses and returned echoes which, when studied on an oscilloscope, are seen to exhibit subtle differences in frequency which are caused by the relative motion of the target and the radar device. Magellan's radar system is of this kind.

The same antenna which sends the radar pulse also receives the returned signal, and in order to avoid interference between the transmitted and returned beams, illumination of the surface is accomplished by sending discrete pulses just a few microseconds in length, separated by pauses. This is achieved by an electronic switching mechanism which blocks the receiver during transmission periods and prevents interference between the transmitted and received pulses. Then, during each hiatus the returned beam can be received and its contained information stored within a computer. The returned pulse also may be displayed as a line sweep on a cathode ray tube and recorded on film.

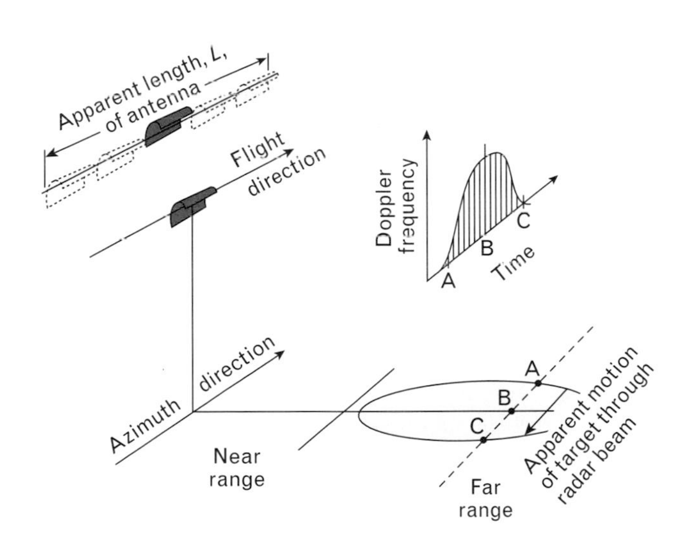

Principles of the synthetic aperture radar system. The antenna is relatively small but transmits a moderately broad beam. The record of Doppler frequency changes as the target moves through the repeated radar pulse, enabling the target to be resolved on the film as though the antenna was of length *L*.

downward and to the side of the antenna, for which reason it is known as *Side-Looking Radar* or SLR, for short. Because this is how radar images are obtained, radar pictures of large regions of planetary surfaces generally are put together from several individual image strips.

Because every radar pulse is electromagnetic radiation, its velocity of transmission is known; it travels at the speed of light in a vacuum. Since it is possible with modern technology to time very accurately the period which elapses between the transmission of a radar pulse and the time of return of the echo, radar may be used to undertake topographic mapping. Therefore, to complement the wide-angle side-looking antenna aboard the Magellan spacecraft, there was a second, narrow-beam, antenna which points more-or-less vertically downwards. This provides topographic transects of the Venusian surface.

Interpreting radar images

Radar images are trickier to interpret than visual ones. Thus, although radar pulses cast shadows when they encounter obstacles, highly-reflective smooth surfaces that would show as bright regions in a traditional photograph may not appear bright on a radar image; indeed, they might appear black. This is highly confusing to the non-expert. You see, a radar beam hitting a highly reflective (e.g. a metallic) surface, rather than being reflected back to the receiver, is reflected away from it, according to the well-known law which states that the angle of incidence is equal to the angle of reflection. This being so, no echo is returned to the antenna. On the radar image, therefore, the metallic region would appear black, like a shadow. Because the radar beam hits the target surface at gradually more oblique angles along the image strip in the look direction, it is necessary to take account of incidence angle when studying the returned images, as it affects the nature of the return pulse.

The radar return is also a function of several terrain properties, whereby the transmitted radar pulse interacts with the surface with which it comes in contact and is modified before being sent back to the receiver. Factors which conspire to modify the beam include the nature of the surface rock, its roughness on the centimetre scale, the slope of the ground and the surface's electrical properties. With all these factors having to be borne in mind, not surprisingly the interpretation of radar images is significantly more difficult than that of visual imagery and demands considerable specialist experience.

The specific kinds of radar echoes returned from Venus indicate that there are radar-bright areas which represent a high degree of backscattering due to rough surface materials, and radar-dark areas which generally represent smooth surfaces. It is therefore possible to gain some idea about the types of rocks which outcrop from one point to the next, and also of the surface texture. When this information is coupled with altimetric data, scientists can begin to unfold the planet's geological history.

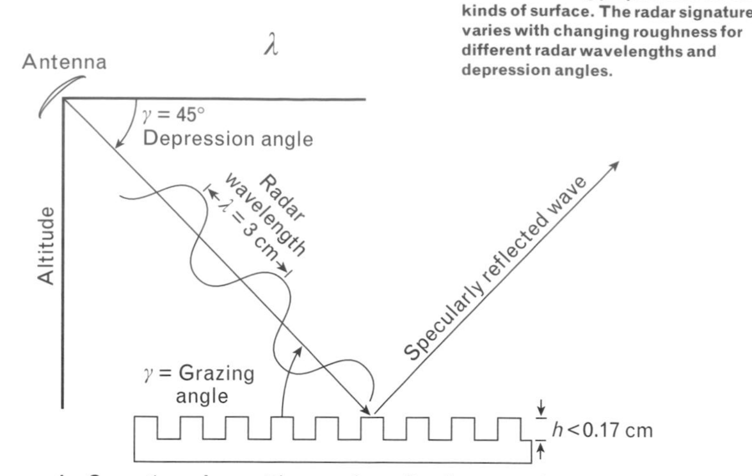

A Smooth surface with specular reflection; no return

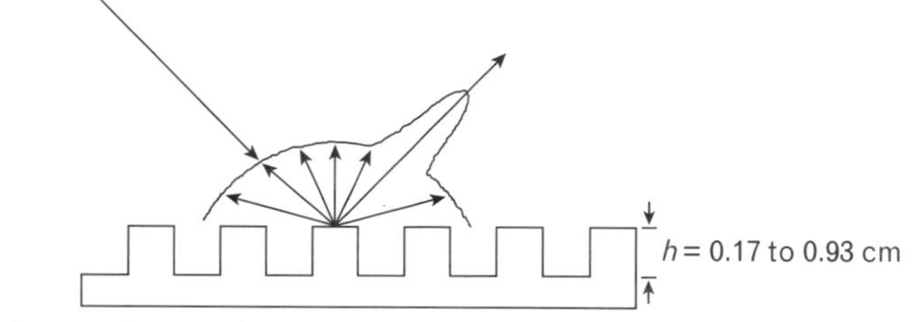

B Intermediate surface roughness; moderate return

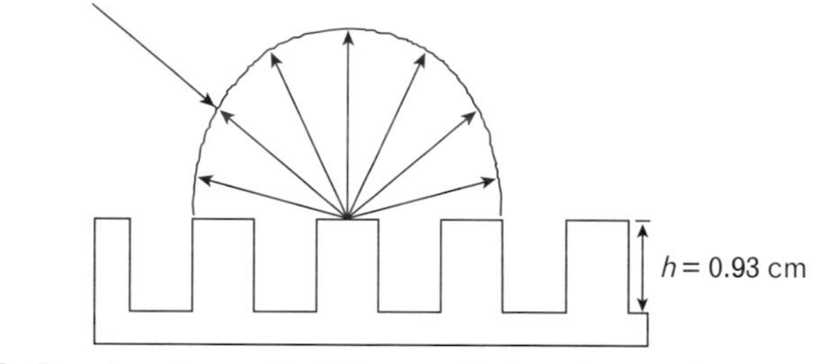

C Rough surface with diffuse scattering; strong return

Radar scattering properties of different kinds of surface. The radar signature varies with changing roughness for different radar wavelengths and depression angles.

Earth-based studies

Radar observations have been made from the Earth for over two decades; indeed, they gave us our first information about the planet's surface. Such observations have been made principally from three US locations: the 43-m antenna at Haystack in Massachusetts which operates at 3.8 cm wavelength; the 64-m antennae based in the Californian Mojave Desert at Goldstone, operating principally at 12.5 cm; and the 300-m system at Arecibo in Puerto Rico, which operates mainly at 70 cm wavelength. Because of the geometry of the Earth with respect to Venus, high-resolution images can be obtained only around inferior conjunction – once every 19 months (see Appendix 3).

The initial work using this technique was undertaken purely for distance measurement and these data were collected during the late 1960s. However, by the year 1970, images of the Venusian surface had been collected by each of the observatories. At this time the best horizontal resolution that could be achieved was around 50 km; thus it could only be shown that Venus had some fixed features and that there was a regional variation in radar brightness. The first actual features to be resolved, as techniques improved, were circular ones believed at the time to be craters. The highest resolution achieved at Goldstone is around 10 km. The most powerful of the three radar telescopes is that at Arecibo, where the best imagery can resolve features as small as 3 km. At the forefront of research here is the team of Campbell and Burns, who were the first to build up a mosaic of the surface which covered about 25% of the globe at a resolution of between 10 and 20 km. The images show considerable detail which mainly reflects differences in roughness from one area to the next. The radar-bright signatures from regions such as Freyja and Akna Montes, bordering Lakshmi Planum, from Maxwell Montes and from the two massifs of Rhea and Theia Montes in Beta Regio, imply rough topography at these radar wavelengths. Several large circular structures with radar-bright rims are seen, together with linear zones of banded terrain, seen only in the highest resolution imagery, and strongly suggestive of folded rocks. The existence of global-scale belts of radar-bright terrain, in places cross-cut by

Arecibo radar image of Beta Regio. The distinctive radar-bright signatures of the volcanic shields, rift fault walls and tessera massifs are evident. Several impact craters also can be seen.

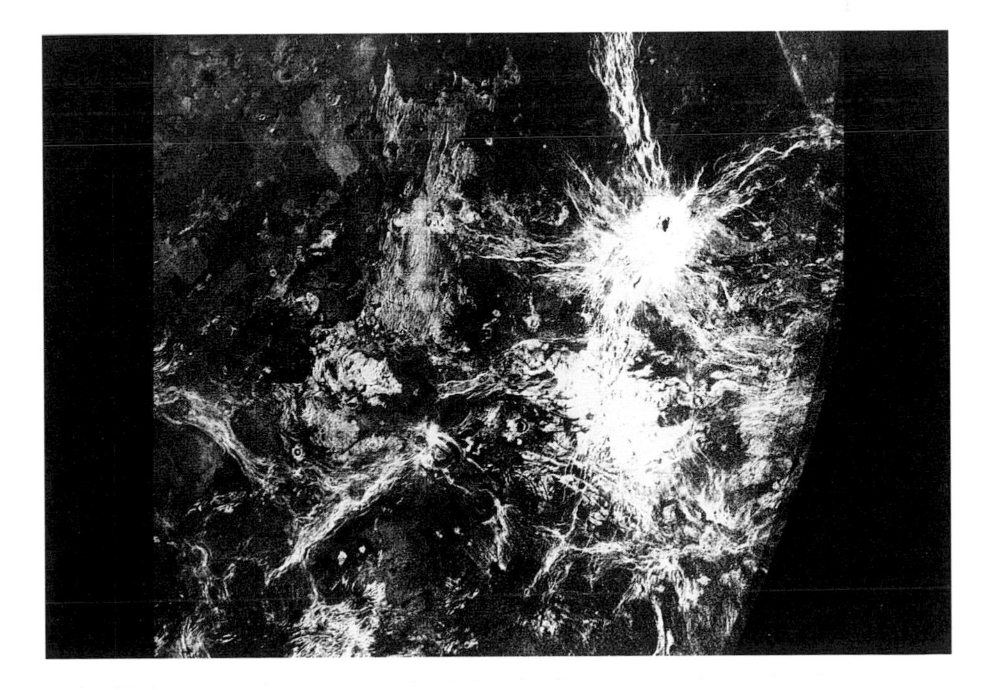

linear fault zones, was clearly established by the radar mapping undertaken from Puerto Rico.

Pioneer-Venus

In terms of planetary mapping capabilities, the first landmark was the successful deployment of the US spacecraft, Pioneer-Venus 1 and 2, in December 1978, an achievement which was to cost NASA more than 175 million dollars. The latter craft, which arrived five days after the former, comprised five separate craft – first, a hardware carrier known as a "bus", secondly, a hefty atmospheric probe and lastly, three smaller probes which entered the Venusian atmosphere at different points and determined the structure and composition of the air on both the day- and night-sides. At much the same time, Pioneer-Venus 1 was placed in orbit; between them, twelve different scientific experiments were included in their payloads. Interestingly, although the periapsis of the Orbiter orbit was originally 378 km, by 1986 it had risen to 2200 km, after which it has gradually diminished; it had fallen to 150 km by 1992.

The Pioneer-Venus 1 Orbiter carried a varied payload, including instruments to study the surface of the planet, its interaction with the solar wind, the magnetic field, plasma waves and the atmosphere. It also carried a radar mapping package which included a modest-resolution mapper and a radar altimeter which allowed the mission team eventually to produce the first accurate topographic map of Venus. It effectively stalked around the planet once every 243 days. By the end of its scheduled mission, Pioneer-Venus 1 had mapped about 93% of the surface, between latitudes 74°N and 63°S, and had obtained altimetry with a vertical accuracy of about 200 m, integrated over an area 100 km square. It showed Venus to be pretty flat over large areas, but it also showed that the landscape was chacterized by extensive regions of *upland rolling plains*, less extensive areas of *lowlands* and much more restricted areas of *highlands*. The altimetric data and mapper data were combined to give the first general relief map of the planet.

Of the lowland regions mapped by Pioneer-Venus, one – Guinevere Planitia – was particularly prominent and it, together with adjacent lowlands in the equatorial belt, formed a somewhat rectilinear feature shaped like a huge cross. The largest lowland, Atalanta Planitia, lies generally 1.4 km below the mean datum level and has roughly the area of the Gulf of Mexico. Its smoothness and general lack of features resembling impact craters suggested to many workers that it was similar to the lunar maria. Gravity data suggested that the lowland plains were regions of thinner crust and lower density than the upland plains.

The highlands stand out clearly on the Pioneer-Venus topographic map, and occur in three principal areas: (i) Ishtar Terra, (ii) Aphrodite Terra, and (iii) Beta Regio. Ishtar is about the size of Australia and Aphrodite as large as Africa. The highest point on the planet's surface resides in Ishtar Terra, where a point in Maxwell Montes rises to 11 km above datum. Immediately to its west lies a vast high plateau, named Lakshmi Planum, which is bounded by narrow zones of compressional landforms.

No point in Aphrodite Terra rises as high as Maxwell Montes, whose mean height is around 5 km. The highest region lies in the

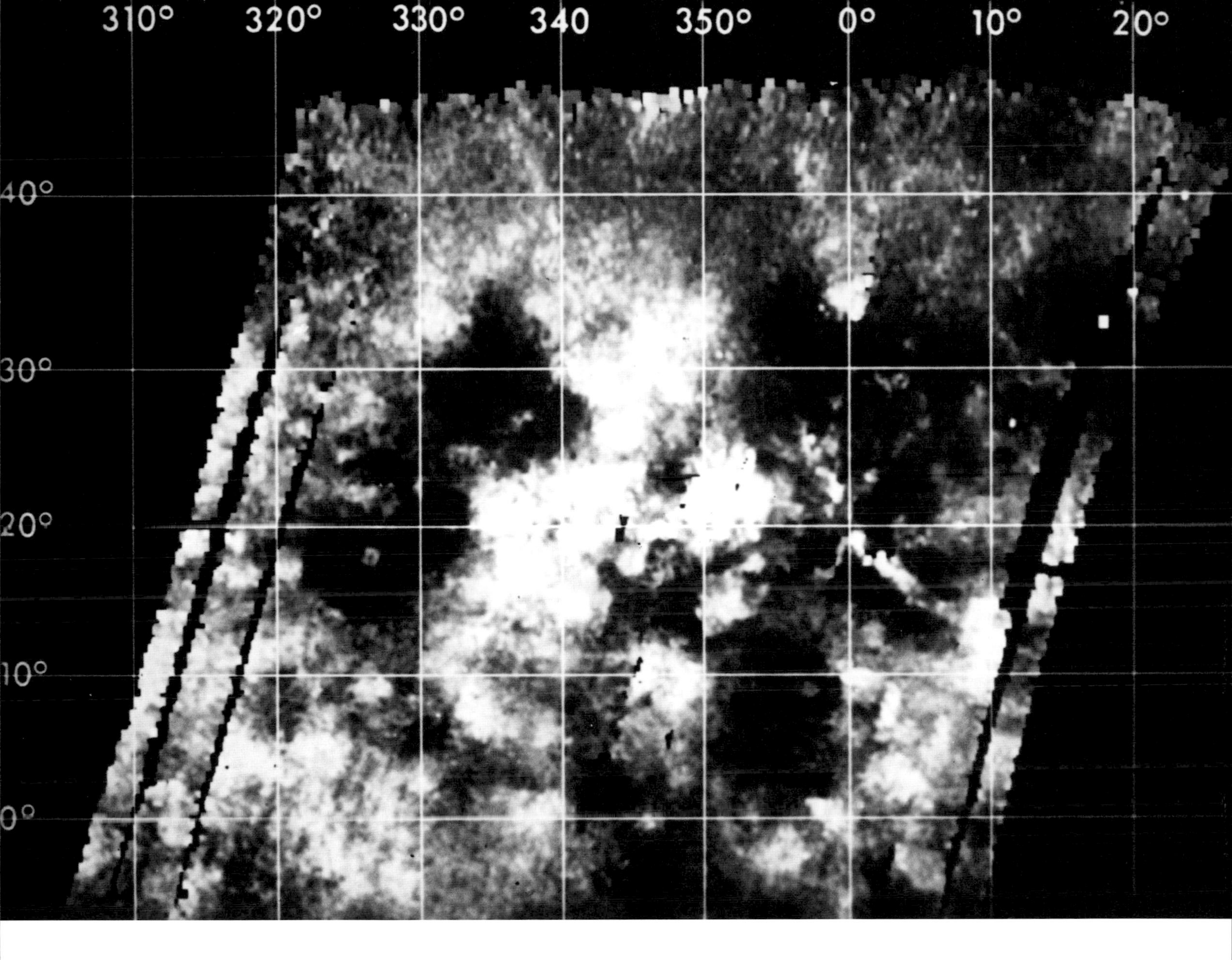

Pioneer-Venus mosaic of western Eistla Regio. The most prominent features are the radar-bright volcanoes of Sif and Gula Montes, with their associated lava flows.

west, while in the east, a curved mountainous belt gives rise to the region called Atla Regio. While exhibiting similar linear ridge-and-trough features to those found on Ishtar, Aphrodite has a more degraded look which was interpreted by some scientists to mean a greater age. To the south is a deep curvilinear chasm – Artemis Chasma – now believed to be an enormous corona structure. Linear valleys cross the central parts in an approximately WSW/ ENE direction, one being at least 1400 km long and plunging 7 km below mean datum, which firmly places it in the same league as Valles Marineris of Mars.

Pioneer-Venus also revealed a range of shallow circular features between 200 and 700 km in diameter which were interpreted as either impact or volcanic structures. One very prominent radar-bright circular feature, originally named Sappho (7°N 15°E), has a diameter of 250 km and was surrounded by radiating radar-bright features, believed to be lava flows. Similar radar-bright structures were found in Beta Regio. Subsequent work has confirmed their volcanic nature but also has shown that other quasi-circular radar-bright features are more likely to be impact structures.

Pioneer-Venus eventually suffered its demise on 8 October 1992, after nearly 14 years of travelling around Venus. This probably occurred because it exhausted its stock of hydrazine fuel, used to restore its orbit,

Both the Pioneer and Venera spacecraft made a variety of atmospheric measurements. Thus, the zonal flow within the atmosphere was mapped, and its maximum opacity was found to lie 50 km above the surface. Some dark markings observed through ultraviolet filters, both from Earth and from Mariner 10, were found to be due to sulphur particles within the lower atmosphere, while the upper cloud layers, generally brighter, were interpreted as due to sulphuric acid droplets. One important finding concerns levels of the inert gas ^{36}Ar. This gas was most abundant when the solar system first formed, but since then large amounts of radiogenic potassium ^{40}K have decayed, releasing the isotope ^{40}Ar. As a consequence, the ratio of original $^{36}Ar/^{40}Ar$ gives vital clues relating to planetary evolution. The values obtained by both craft indicate that Venus has much more ^{36}Ar than the Earth. This implies that Venus retained more gas from the primaeval solar nebula than did Earth, which came as something of a surprise to atmospheric scientists, who had anticipated the reverse due to the stripping away of this early atmosphere by the outward-streaming solar wind.

causing the orbit to decay. The result of a loss of elevation would be severe overheating as the satellite dropped into the denser levels of the Venusian atmosphere.

Veneras 15 and 16

The stage was now set for a new wave of exploration: from the time of their arrival in October 1983, until July 1984, Veneras 15 and 16 imaged Venus from elliptical orbits with periapsis of 1000 km and apapsis of 65,000 km. Venus was imaged by their SLR antennae, using the probes alternately at a wavelength of 8 cm. The radar survey cycle began when each northward-travelling orbiter reached 80°N, continued over the polar regions and down to between latitudes 30°–35°N. In this way strip images of regions 8000 km long and between 120 km and 160 km wide were put together to form a large composite image. While the SLR antennae were working, so was the narrow-beam altimetric antenna, acquiring height data with a vertical precision of 50 m, but over an area of between 40 km and 50 km, meaning that reliable surface relief data could only be obtained for features which were at least this size.

The greater resolution of these spacecraft revealed all manner of new landforms, including finer details of ridge- and trough belts not discerned from Pioneer data, complex *parquet terrain*, faults, and a number of impact craters, volcanic calderae and weird *arachnoids*. In one area of 85×10^6 km², 140 craters ranging in diameter from 10 km to 150 km were shown. About a quarter of these had radar-bright haloes, which were believed to represent impact ejecta.

Since the northern polar region of Venus was not imaged by Pioneer-Venus and is inaccessible to Earth-based radar, the polar-orbiting Venera spacecraft provided new data. It turned out to be an extensive plain lying near the planet's average elevation. The plain is not, however, without interest since it is crossed by belts of linear ridges and furrows several hundreds of kilometres long and up to 150 km wide. The most prominent of these runs from Atalanta Planitia, through the poles, and then turns south towards the highlands of Ishtar Terra.

The higher-resolution imagery obtained by Veneras 15 and 16 showed that most of the surface of Lakshmi Planum was relatively smooth and between 3 and 5 km higher than the adjacent area. It identified two prominent circular depressions – Collette, which measures 120 km × 80 km, and Sacajawea, which is larger (280 km × 140 km). They appeared to be volcanic calderae.

Structure of the atmosphere of Venus.

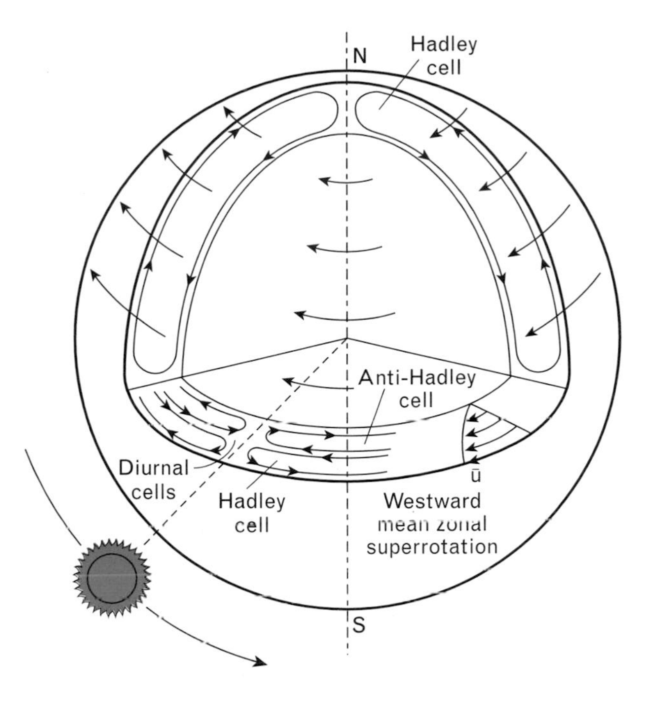

Sketch of possible circulation patterns in the atmosphere of Venus. The mean zonal velocity $\bar{u}$ is a westward superrotation. The magnitude of $\bar{u}$ increases with height above the surface. The meridional Hadley circulation may not extend to the poles. Centers of convergence and divergence in the diurnal Hadley and anti-Hadley circulations may not occur at noon. The diurnal circulations have other flow components not shown here.

Lakshmi Planum was found to be almost completely surrounded by belts of linear ridges and grooves, its borders being defined by the high ranges of Akna Montes to the west, Freyja Montes to the north and Maxwell Montes in the east, while a narrow belt of less prominent ridges marks the other margin which slopes down steeply to a feature called Vesta Rupes. These prominent mountain massifs are the highest on Venus, being 8 km, 9 km and 12 km above datum respectively. Workers who have studied the images of the ridges say that they look rather like stacks of giant plates, suggesting that they could be either folded rocks or complex fractured slices of crust. At the eastern end of the plateau small regions are traversed by intersecting, closely spaced ridges and grooves which are often embayed by the plateau surface materials. These relationships seem to suggest that the ridge-and-groove landforms may be remnants of whatever lies beneath the surface skin.

The intersection of ridges and grooves is seen elsewhere on Venus and because of the distinctive pattern – it resembles wood-block flooring – it was originally called parquet terrain but is now more frequently described by the word *tessera*. It is particularly well seen to the east of Maxwell Montes in a zone measuring 1000 × 2000 km.

East of Ishtar is a region which is predominantly a plain but which has several upland massifs. Venera images reveal that this region – Tethus Regio – has areas covered with groups of dome-shaped hills and also several 300–500 km diameter elliptical structures. The latter typically have a core of somewhat irregular relief which may be surrounded by up to 12 rows of concentrically arranged ridges and grooves. Soviet workers coined the term *coronae* for these. At the time of their discovery, ideas concerning the origin of these peculiar landforms were

Veneras 15 and 16 mosaic of the northern region of Venus between longitude 0° and 60°N. Maxwell Montes lie on the left-hand side, with the circular structure, Cleopatra Patera, being clearly visible. To the east of this lies an extensive area of parquet terrain, now called tessera.

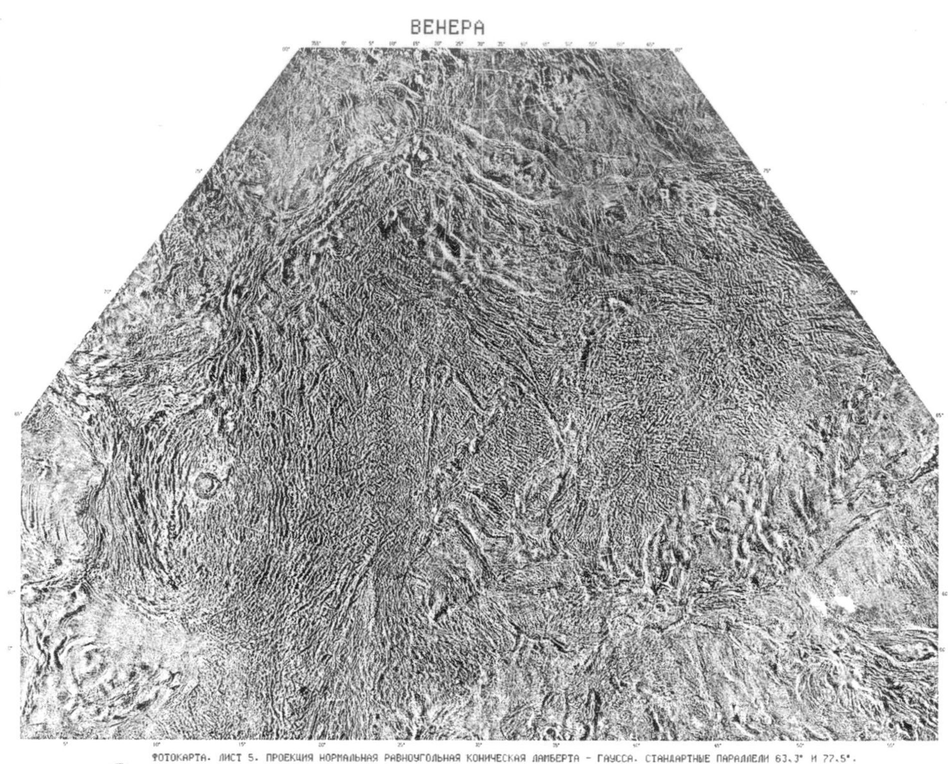

Veneras 15 and 16 mosaic of ridge-belts and corona structures in the regions of Bereghinya and Leda Planitiae.

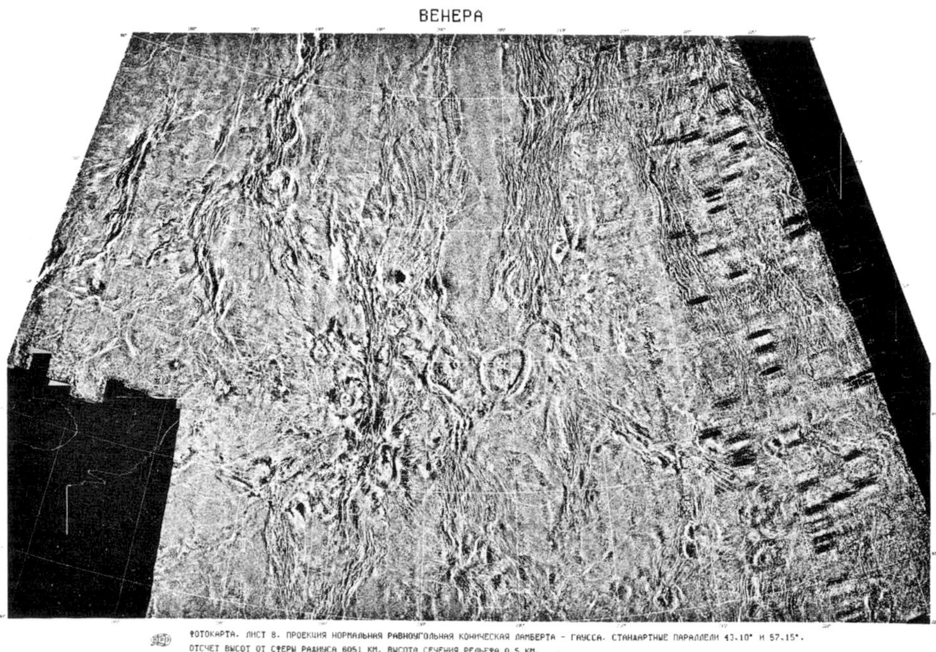

considered entirely speculative, although radar-bright flow-like markings surrounding some coronae were thought to represent lava flows. Furthermore their concentration, especially at the western and eastern margins of Ishtar Terra, suggested some underlying structural control. Since volcanism and tectonism usually go hand-in-glove, the most favoured explanation was that they had an endogenic origin and were a manifestation of peculiarly Venusian volcanic processes.

Southwest of Ishtar Terra lies Beta Regio, an upland region dominated by two prominent 4.5-km-high mountains, Theia and Rhea Montes. Venera images revealed that both of these have low flank slopes and show radar-bright radiating streaks which are presumed to represent lava flows. Veneras 9 and 10 both landed on the edge of Beta Regio and showed the surface rocks to have a basalt-like chemistry.

Only the northern part of this region was surveyed by the last two Veneras, 15 and 16, which showed the transition from the adjacent plains to be gradual rather than marked. In passing upslope, numbers of linear and often arcuate scarps appeared which were hundreds of kilometres long and between 5 km and 10 km in width. In places these were found to pass into fault-bounded troughs called *graben*. The association of volcanic structures and faults is quite marked and bears close resemblance to the relationships they show on both the Earth and Mars.

THE MAGELLAN MISSION

6

The most recent radar mapping mission was that of Magellan which completed four 243-day-long mapping cycles that provided planetary scientists with high-resolution imagery, radiometric, altimetric and gravity data. Our understanding of Venus has improved immeasurably as a result of it, and new maps of the planet are published here.

During the early 1970s, NASA drew up plans for a very sophisticated mission to Venus. Named Venus Orbiting Imaging Radar, or VOIR for short, this was to have a circular orbit around Venus and a complex array of experiments. However, after detailed studies by technicians and scientists, the mission proved too costly to be viable and was scrapped 12 years later. Then in late 1983, the Russian Veneras 15 and 16 spacecraft reached Venus and provided excellent medium-resolution imagery of selected parts of the planet, using a wavelength of 8cm. They revealed a host of hitherto unsuspected landforms, including coronae, arachnoids, ridge-belts and many impact craters and made a major contribution to the first global geological map of Venus.

Despite the undoubted success of Venera, the pressure for another mission did not go away, and in October 1983 a much-reduced NASA mission objective saw the inception of the Venus Radar Mapper, or VRM programme. Its objectives were to obtain high-resolution radar images, altimetric data, radiometric information and gravity-field measurements of the planet. VRM was to have an elliptical polar orbit and, to keep costs low, was to make use of spare parts from other missions. In 1985 VRM officially was renamed Magellan. The justification for its launch was that high-resolution imagery and gravity data were required to resolve several hitherto unanswered geological questions. Amongst these was the necessity to understand whether or not Venus showed any evidence for past surface water, had developed Earth-style plate tectonics, and whether or not it was volcanically and tectonically active at the present time.

Magellan was scheduled as the first spacecraft to be launched by the Space Shuttle, and a launch date of May 1988 was fixed. Regrettably this plan was foiled by the 1986 *Challenger* disaster, which set back the launch schedules of all missions in the wake of a 32-month suspension of Shuttle activity. Furthermore, because of planetary configurations in that period, it meant that the Galileo mission to Jupiter now had to go in October 1989 (the planned new launch date for Magellan) and eventually it was decided that Magellan should be launched in May 1989. A new trajectory had to be planned, whereby Magellan had to spend 13 months in space and would go one-and-a-half times around the Sun before arriving at Venus.

The alignment of Earth and Venus during 1989 meant that Magellan had to be launched between 28 April and 28 May 1989. After one false start on 28 April, the Space Shuttle *Atlantis* eventually took the 3433-kilogram spacecraft into the sky on 4 May and, fifteen months later, the Magellan craft went into a highly elliptical orbit around Venus. This crucial manoeuvre happened on 10 August 1990 whereupon, as if by magic, a number of enigmatic problems arose such that it seemed extremely likely the mission might be a complete failure. However, the problems eventually were overcome by NASA's tenacious technicians, and mapping commenced in early September, one month late.

The Magellan spacecraft

The Magellan radar mapper was built by the Martin-Marietta organization, in Denver, Colorado. By and large it was assembled from bits and pieces left over from earlier missions, such as Ulysses, Voyager, Viking and Galileo. The probe, which weighed 3.45 tonnes, carried four antennae: a 3.7-metre-diameter parabolic high-gain antenna (HGA) operating at a wavelength of 12.6 cm, medium- and low-gain antennae (MGA and LGA) and a horn-shaped altimetric antenna (ALTA). It also carried a radiometer, large solar panels which provided 1200 watts of electrical power, and a Starr 488 solid fuel rocket motor which put the probe into Venus orbit. The 3.5-metre-square solar panels collected solar energy for charging the spacecraft's two nickel–cadmium batteries and always pointed towards the Sun. Precise control of the probe was achieved by three momentum wheels, which acted as gyroscopes, maintaining the correct orientation of Magellan relative to Venus. Where necessary, these were assisted by twelve small gas thrusters which were fired to rotate the craft and release the excess speed that built up in the momentum wheels.

The large antennae not only operated as a side-looking radar but also were used for communications purposes. The small horn-shaped antenna pointed directly beneath Magellan, intersecting the surface at right angles and providing altimetric information for each swath of surface it passed over. By the time three mapping cycles were completed – each being one Venusian year long, i.e. 243 Earth-days – 99% of the surface had been mapped. The last mapping cycle saw the spacecraft's periapsis lowered so that the movements of the probe could be monitored very accurately by tracking minute variations in the strength of its radio signals. This final phase provided high-resolution gravity data which are related to mass variations within the planet. Data were collected until October 1994; then, in the second week of October, Magellan burned up in Venus's dense atmosphere. Contact was finally lost on 17 October 1995.

Mapping Venus with Magellan

Using its array of data-collecting instruments, Magellan completed a total of 1852 data-collection and playback passes of the planet during the first 243-day mapping

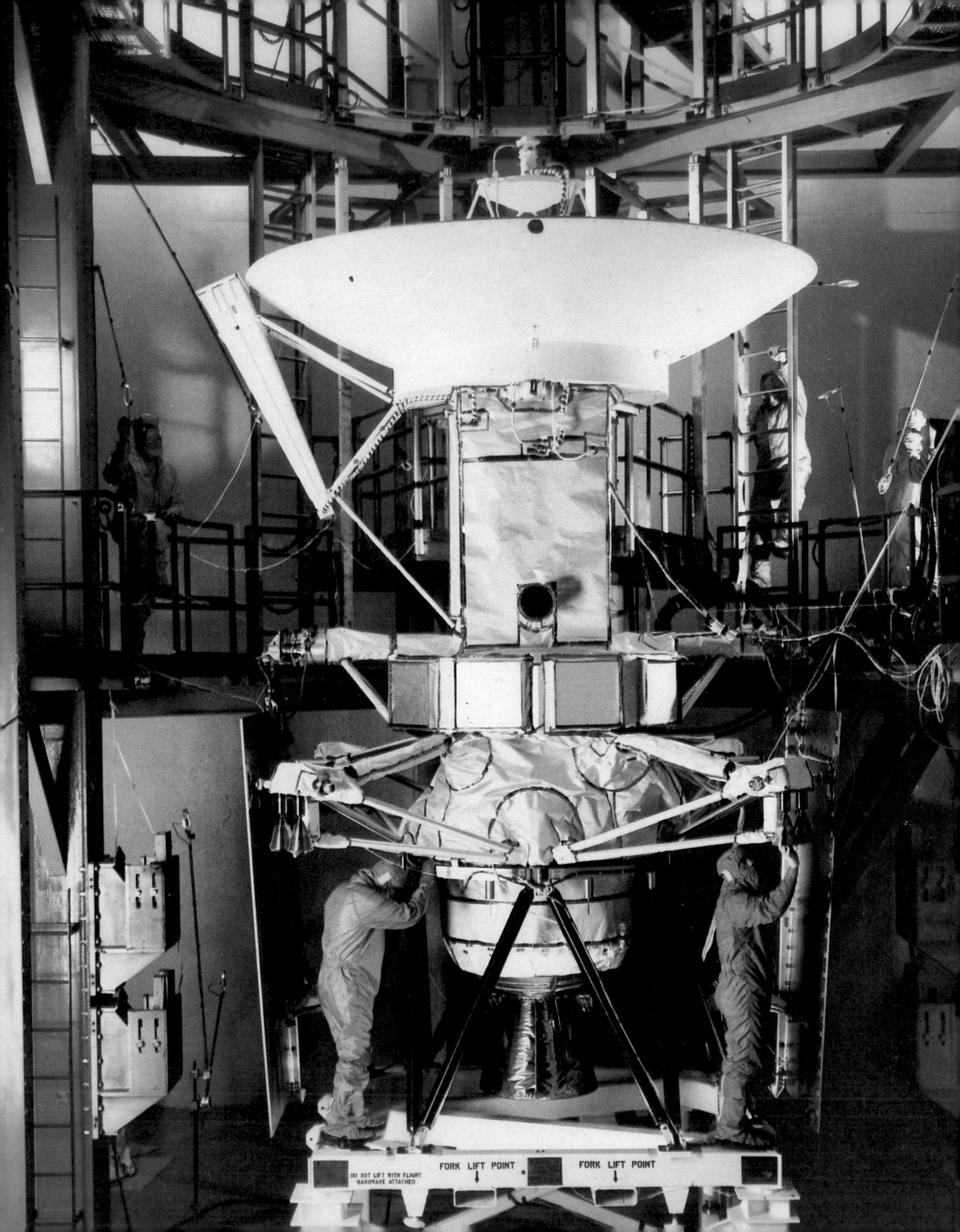
DO NOT LIFT WITH FLIGHT
HARDWARE ATTACHED
FORK LIFT POINT
FORK LIFT POINT

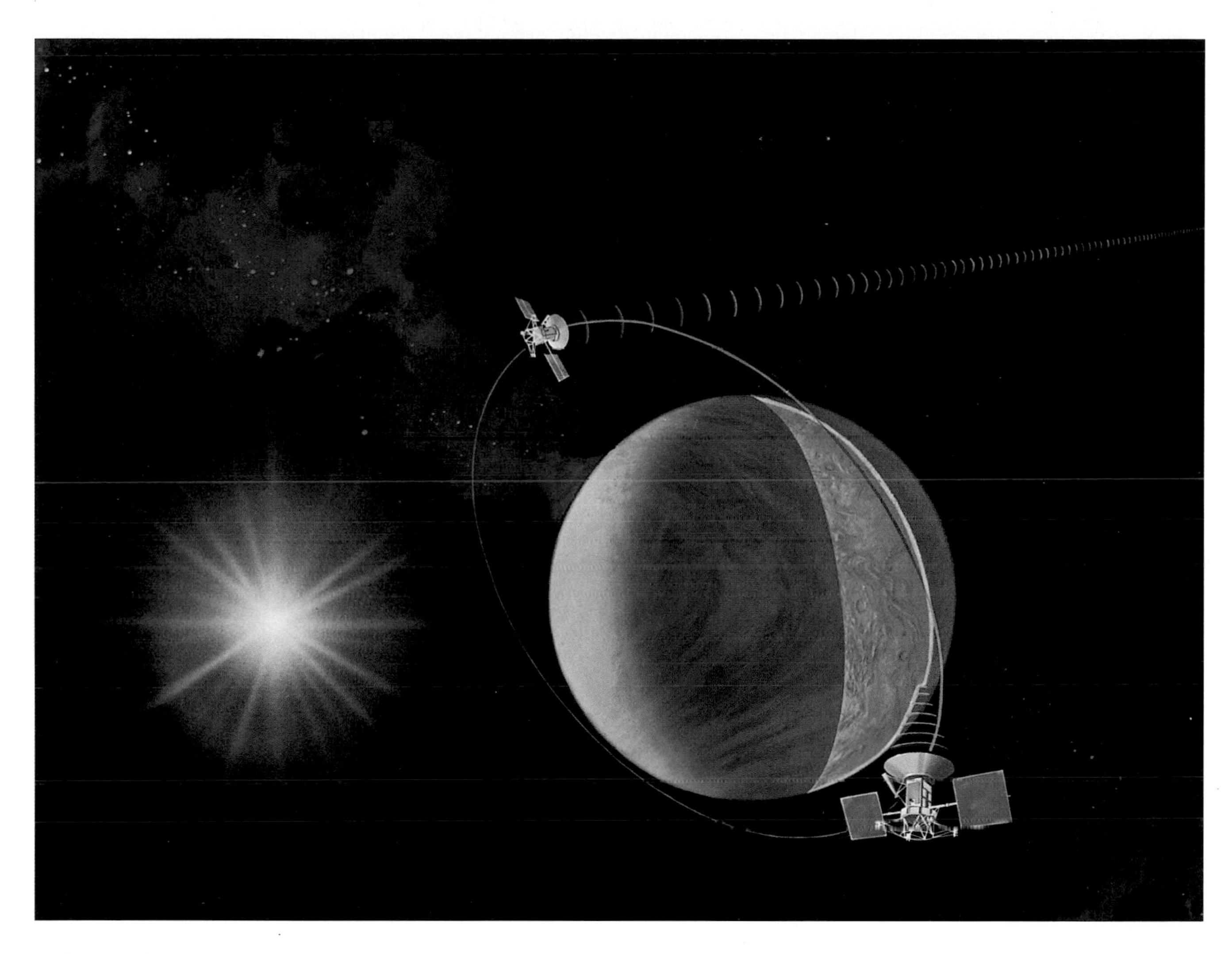

Configuration of Magellan radar mapping probe.

The Magellan spacecraft being prepared for testing at Denver, at the Martin-Marietta Astronautics laboratory.

cycle. Each of the mapping passes collected images over a swath of surface about 25 km wide and 10,000 km long. Magellan also mapped a new area of the planet at each pass, as the planet rotated slowly beneath the craft (which remained essentially fixed in inertial space). Because of the orbital configuration, the HGA had to point to the left as the first mapping cycle proceeded and to the right during the second. During the third, stereoscopic images were obtained.

The long, narrow strips that Magellan imaged were obtained as the probe rotated such that the HGA pointed towards the surface. As the spacecraft altitude fell below 1900 km, the side-looking radar system was activated, the "look" angle of the mapper constantly changing as the probe moved towards periapsis. Mapping ceased when the spacecraft altitude rose above 1900 km, whereupon data storage, communication and idling took place. The mapping pass having been completed, Magellan's HGA was pointed towards Earth, transmitting the imaging data at the rate of 250 kilobits per second. The process was then repeated.

The orbiting radar mapper approached as close as 294 km and receded to 8743 km from Venus during each pass. The imaging system was able to resolve objects as small as about 120 metres. This was a major improvement on previous spacecraft-mounted instruments and also on the powerful Arecibo radar telescope which, prior to Magellan, had provided some of the best images of the planet.

Because of the complexities of a radar system, data could not be collected and

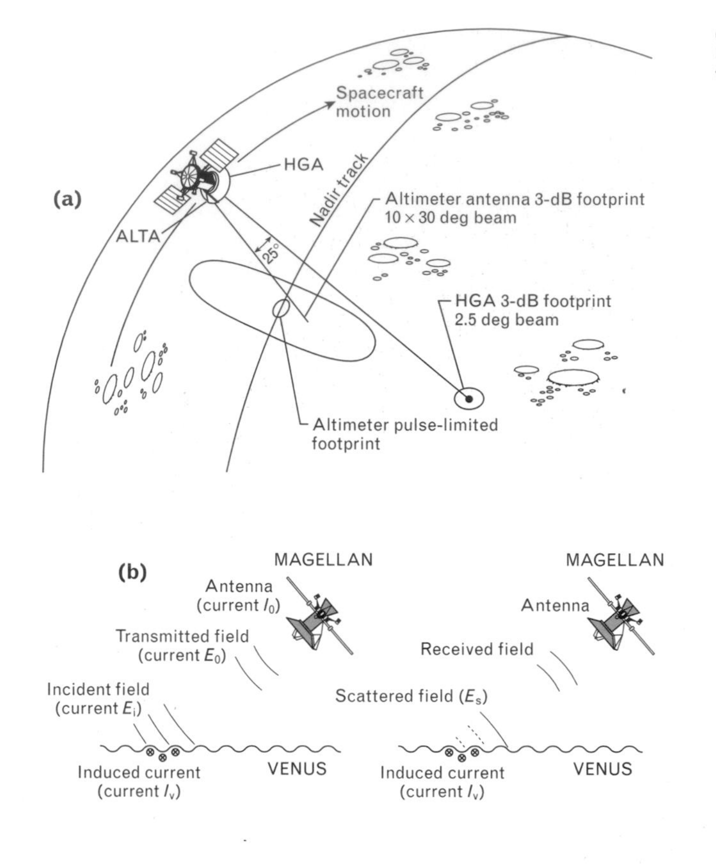

(a) Mapping modes of orbiting Magellan spacecraft. (b) Configuration of transmitted and returned pulse.

transmitted back to receiving stations on Earth at the same time. Thus, each 3 hour 15 minute orbital period had to be divided up into a number of shorter periods for mapping, data playback and idling. Furthermore, on one orbit both high and low latitudes were imaged, while on the next, only the lower latitudes were. By adopting this proceedure NASA avoided duplicating coverage of the polar regions which present a much smaller percentage of new ground during successive passes than do lower latitudes. Thus the probe alternated between mapping from the north pole down to latitude 66.9°S latitude in one orbit, to mapping only from 66.9°N to 74.2°S on the next. During each of the two playback periods allocated for each pass, data were sent back to Earth at the rate of 268.8 kilobits per second, which is only one-third the rate at which new data are recorded. Magellan needed a considerable storage facility!

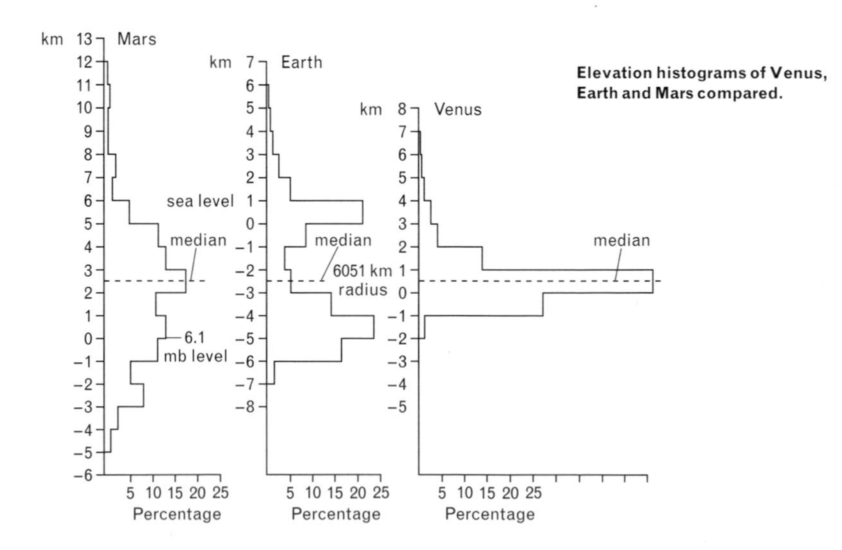

Elevation histograms of Venus, Earth and Mars compared.

Venus and Magellan – the broad view

Magellan confirmed that, in contrast to Earth, where the continents and oceans give a bimodal distribution to the elevation histogram, a similar histogram for Venus is unimodal and the topographic peak narrow. This is also in complete contrast to the histograms for the Moon and Mars, and is a reflection of the extreme average smoothness of the Venusian globe. Amazingly, 90% of the surface lies within a 3 km height interval. Due to the extreme accuracy of Magellan's altimetric data, it was also possible to establish that the steeper slopes (on the metre scale) were characteristic of the more elevated regions. Whereas on Earth large regions have zero slope due to planation of the continental interiors by millenia of weathering and erosion, Venus has only limited zero slope areas. Generally these are occupied by volcanic plains. There is also a broad tendency for the roughest surfaces to coincide with the higher ground, which is why radar-bright regions on Magellan images generally delimit highland massifs or the rims of raised circular landforms such as craters, or the flanks of large steep-sided domes.

Features first revealed by Magellan include wind streaks and aeolian dunes similar to those seen on Mars, lava flow lobes, and different kinds of volcanic channels (some of amazing length). Also impact craters – some as small as 3 km across – with which are associated widespread exposures of radar-bright outflow material associated with the cratering process. The fine details of coronae were also revealed, as were those of ridge-belts and tesserae massifs.

The most widespread terrain types on the planet are lowland volcanic plains. These

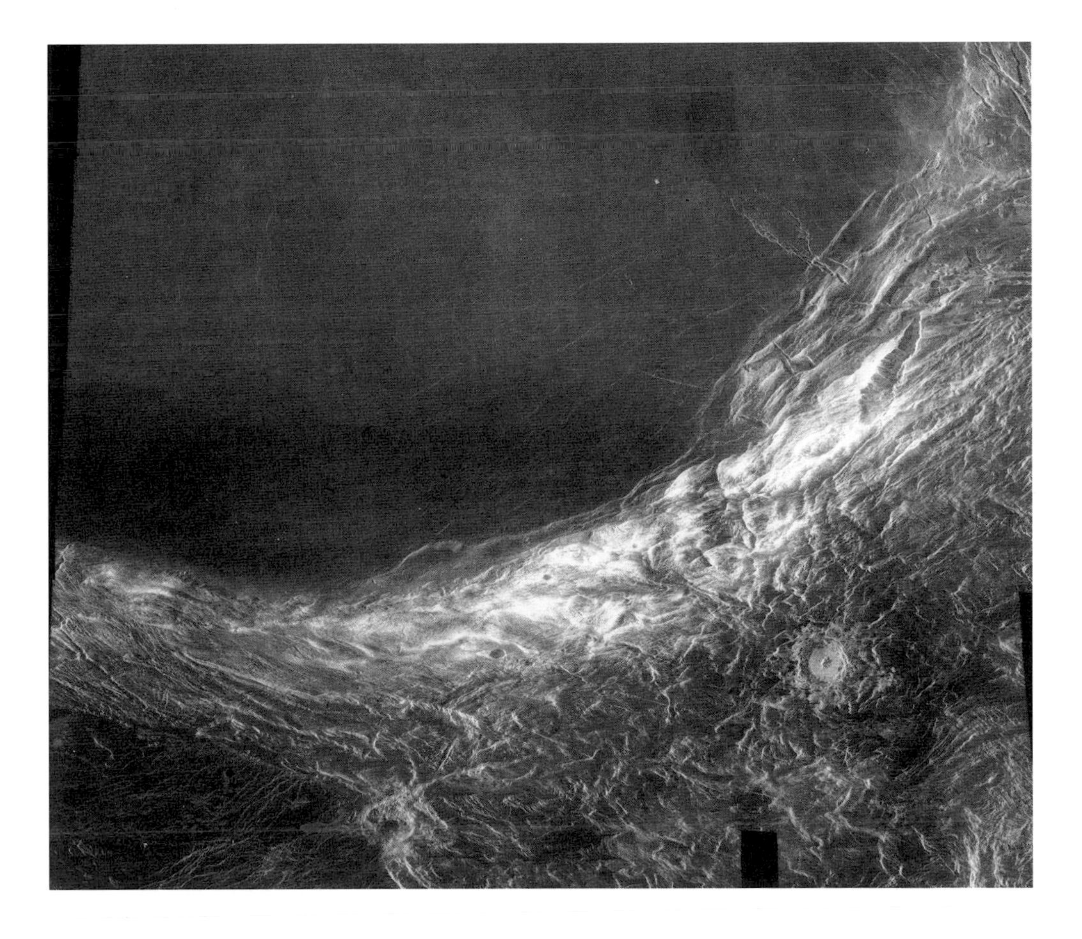

Diverse radar signatures of terrains in the region of Danu Montes and Lakshmi Planum, Ishtar Terra. [Framewidth 460 km.]

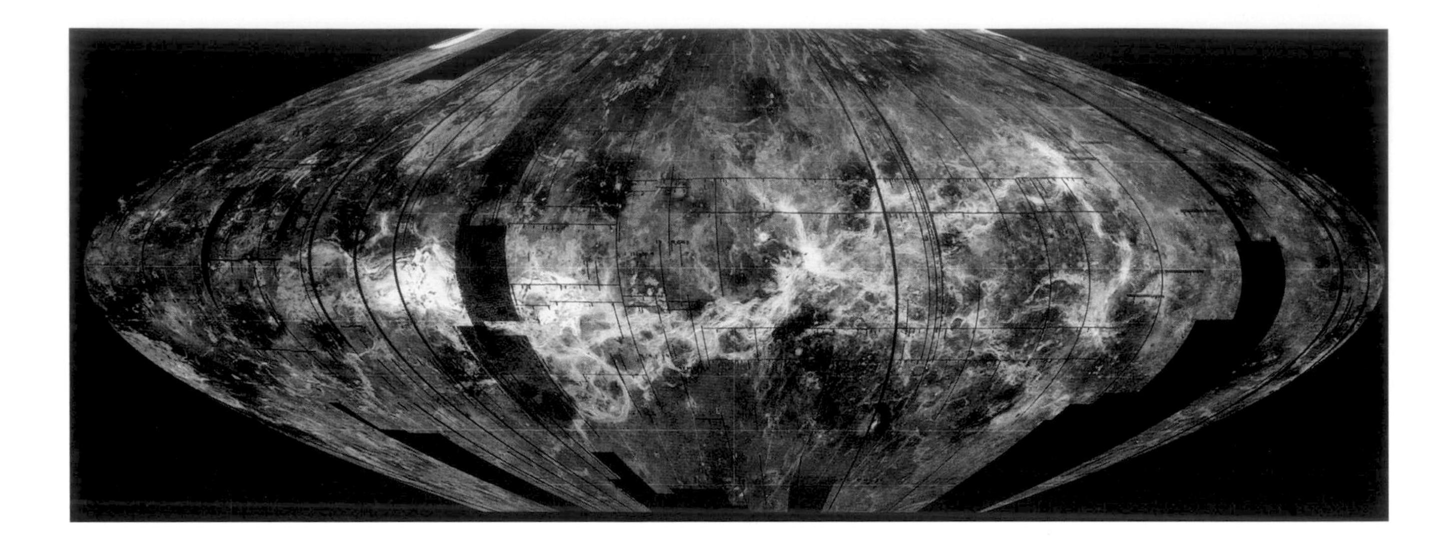

cover 85% of the surface and are host to a plethora of volcanic landforms, many of which were revealed for the first time. The fine detail of these structures is still being analysed by the planetary community. The circular or elliptical volcano–tectonic structures which include coronae, arachnoids and novae, are uniquely Venusian. So also are the numerous fields of small volcanoes called shield fields, and the large, flat-topped pancake domes, thought to be composed of relatively viscous magma. It is now clear that Venus can be thought of in terms of two principal landscape types: volcanic plains – upon which are superposed thousands of individual volcanic landforms – and tectonically-deformed highlands, which cover the remaining 15%.

The slow retrograde motion of Venus meant that the major topographical features were imaged in sequence, as a function of "days after Venus Orbit Insertion". The initial images were of the highland region of Beta Regio, which already had excited geologists after having been beautifully imaged at relatively high resolution by astronomers at Arecibo. Magellan's enhanced resolution imagery showed that in northern Beta Regio – formerly believed to be a huge shield volcano – a large area of complex tessera terrain had been rifted apart by major extensional forces. As each pass took place, it gradually became clear that Venus has had a very active geological history, right up until quite recent times, when a major episode of resurfacing occurred, wiping clean the past history of very large regions.

The structure of the highland regions is often amazingly complicated, and planetary scientists have christened them "CRT" (short for complex ridged terrain). This rather impersonal term is used alongside the word tessera. Built from a complex network of intersecting ridges and troughs, these tend to occur as elevated plateaux with steep sides; they typically are rough at metre and smaller scales. Magellan mapping roughly tripled the known extent of such terrain, it now being recognized in Ovda, Thetis, Phoebe, Beta and Asteria Regiones, and within the regions of Lada Terra and Nokomis Montes. The largest contiguous area of such terrain is found between Ovda Regio and the western extension of Aphrodite Terra. This occupies an area of approximately ten million square kilometres.

Each of the major highland areas is separated from others by deep basins within which are located smaller uplands, such as Alpha, Bell, Eistla and Tellus Regiones.

Global view of surface of Venus – A sinusoidal equal-area projection, centred at 180° east longitude, of all Magellan SAR data gathered from the beginning of the mapping mission to July 1991.

Some of these are joined by deep rifts termed chasmata. Straighter but shallower rifts, arranged in parallel sets, cut obliquely across the equatorial highland zone.

Lithospheric deformation is manifested in global-scale mountain- and ridge-belts of various styles and in the occurrence of large numbers of coronae. In general terms the distribution of these areas of deformation suggests strain over rather broad areas, rather than a restriction to the narrow zones so typical of the Earth's lithospheric plate margins.

The improved resolution of Magellan images has greatly improved our knowledge of the fine structure of the Venusian plains. Details of over 1600 volcanic landforms, of hundreds of coronae, and of complex strain structures and deformational belts were revealed for the first time. The latter are of more than one type and tend to concentrate in or near broad topographic lows, particularly in Lavinia, Atalanta and Vinmara Planitiae.

Of course, Magellan obtained a variety of data. Thus, at the same time as visual images were being studied, other teams of scientists were analysing thermal emissivity data, measurements of dielectric permittivity, and detailed scattering properties. In this way it may become possible to characterize the surface roughness, rock density and (possibly) chemical composition of the Venusian crust. Bearing in mind that Magellan mission scientists are involved in the detailed mapping of an Earth-sized planet, at a resolution equivalent to the best Earth-orbiting probes, it comes as no surprise that there is an unprecedented amount of new information to assimilate. Nevertheless, even now, great forward strides have been taken in understanding this fascinating world.

THE VENUSIAN PLAINS – I

7

Eighty-five percent of Venus is covered in plains which have been formed predominantly by volcanic processes. The plains have also been subjected to tectonic forces. Many of the landforms seen are familiar from terrestrial experience. However, there are several uniquely Venusian features which provide vital clues as to the way in which heat has been lost from the planet's interior.

Eighty-five percent of the surface area of Venus is covered by plains on which are volcanic landforms, craters, aeolian features and linear deformation belts. The global distribution and characteristics of the tectonized zones demonstrate that vertical and horizontal stresses are operating on the Venusian lithosphere in different ways in specific regions. The tally of impact craters – much smaller than that for the lunar highlands or the cratered plateau of Mars – indicates a relatively young age for the presently-

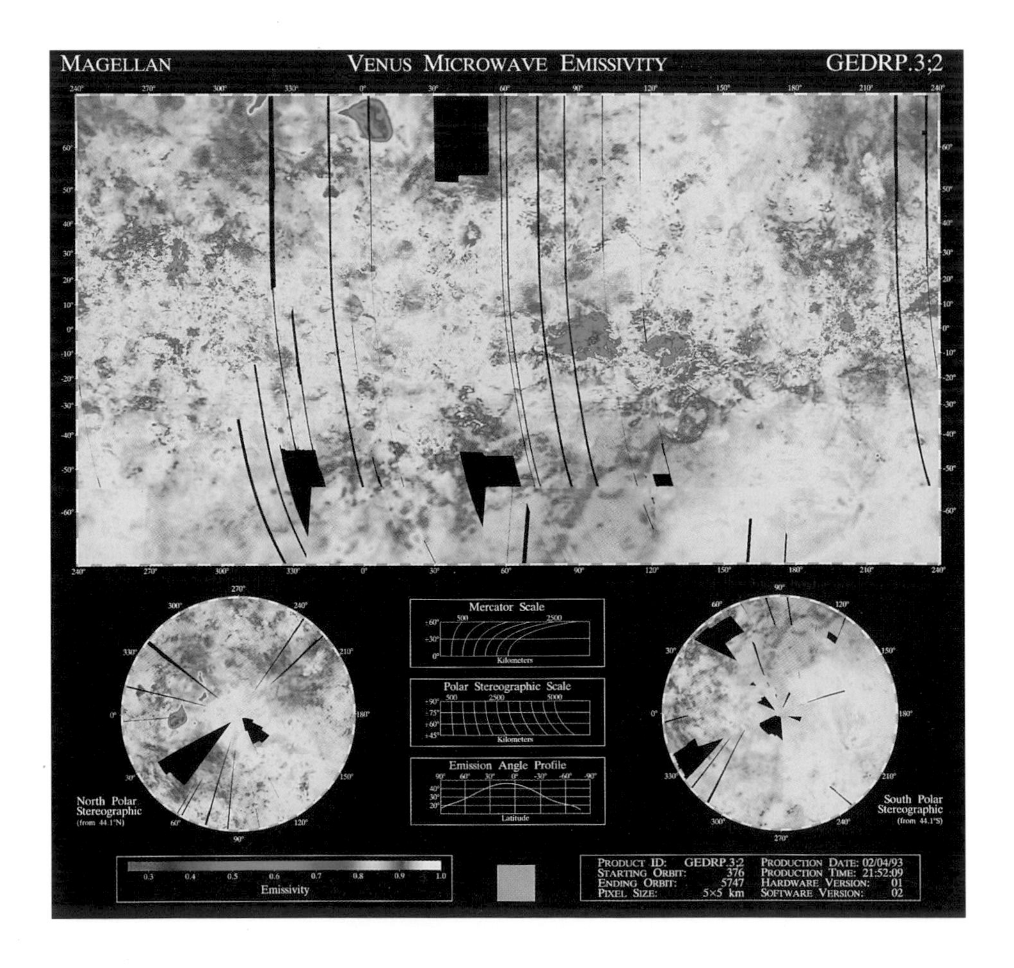

This Magellan microwave emissivity map shows how different regions of the surface radiate heat compared to a perfect radiator. The upland regions, such as Ishtar Terra and Aphrodite Terra show lower values than the lower-lying plains. Thus the colour-coded global map conveniently picks out the highland massifs from the extensive Venusian plains.

exposed surface of the planet. The activity of Venusian winds manifests itself in a variety of features, including wind streaks, crater haloes and dunes.

Types of plains

Two broad kinds of plains terrain had been distinguished prior to Magellan: rolling plains which typically lie between 2 km and zero datum, and lowlands which generally lie below 0 km. The main lowland region, which includes Sedna, Guinevere, Aino, Niobe and Atalanta Planitiae, forms a cross-shaped belt that crosses the equator and whose eastern arms are bisected by Aphrodite Terra. Just to the north of this extensive low-lying zone are found the greatest relative elevation differences on the planet.

Besides elevation differences, rolling plains have radar scattering properties and slopes intermediate between the highlands and lowlands, the latter being between 1.5° and 3° on the metre to decametre scale. Lowlands have darker, smoother radar signatures, and appear to be complex lava-flooded basins. They also host fewer impact craters than the rolling plains and have small gravity lows associated with them.

The improved resolution of Magellan images dramatically enhanced what was known about the plains regions, revealing over 1600 volcanic landforms, large numbers of coronae, and numerous impact craters. Magellan also imaged the details of complex strain structures and deformational belts known as ridge- and fracture-belts which tend to concentrate in or near broad topographic lows. Their distribution contrasts with coronae, which tend to occur in large

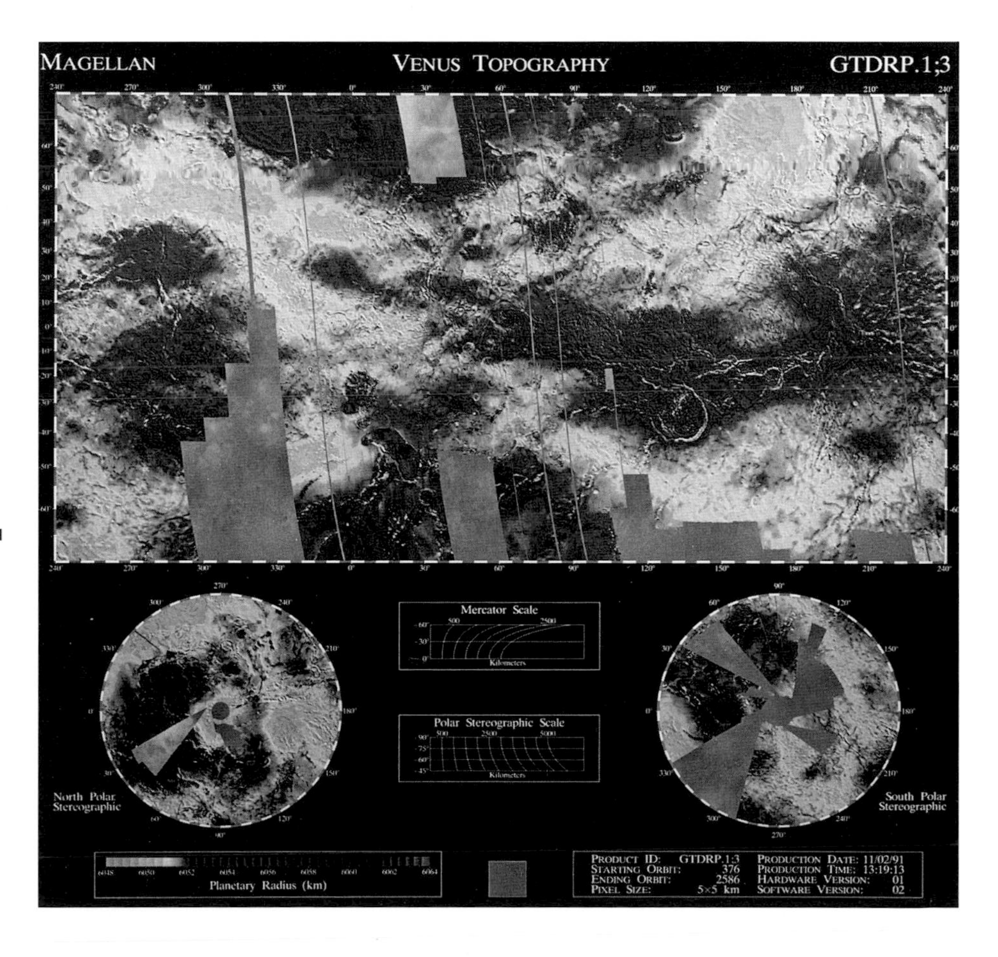

The Venusian topography is illustrated in this Magellan global map and can be compared with the emissivity map shown opposite.

Reticulate fractures cutting radar-dark plains in the region of Guinevere Planitia. At least a dozen volcanic domes, some with small summit pits, can also be seen in the image.

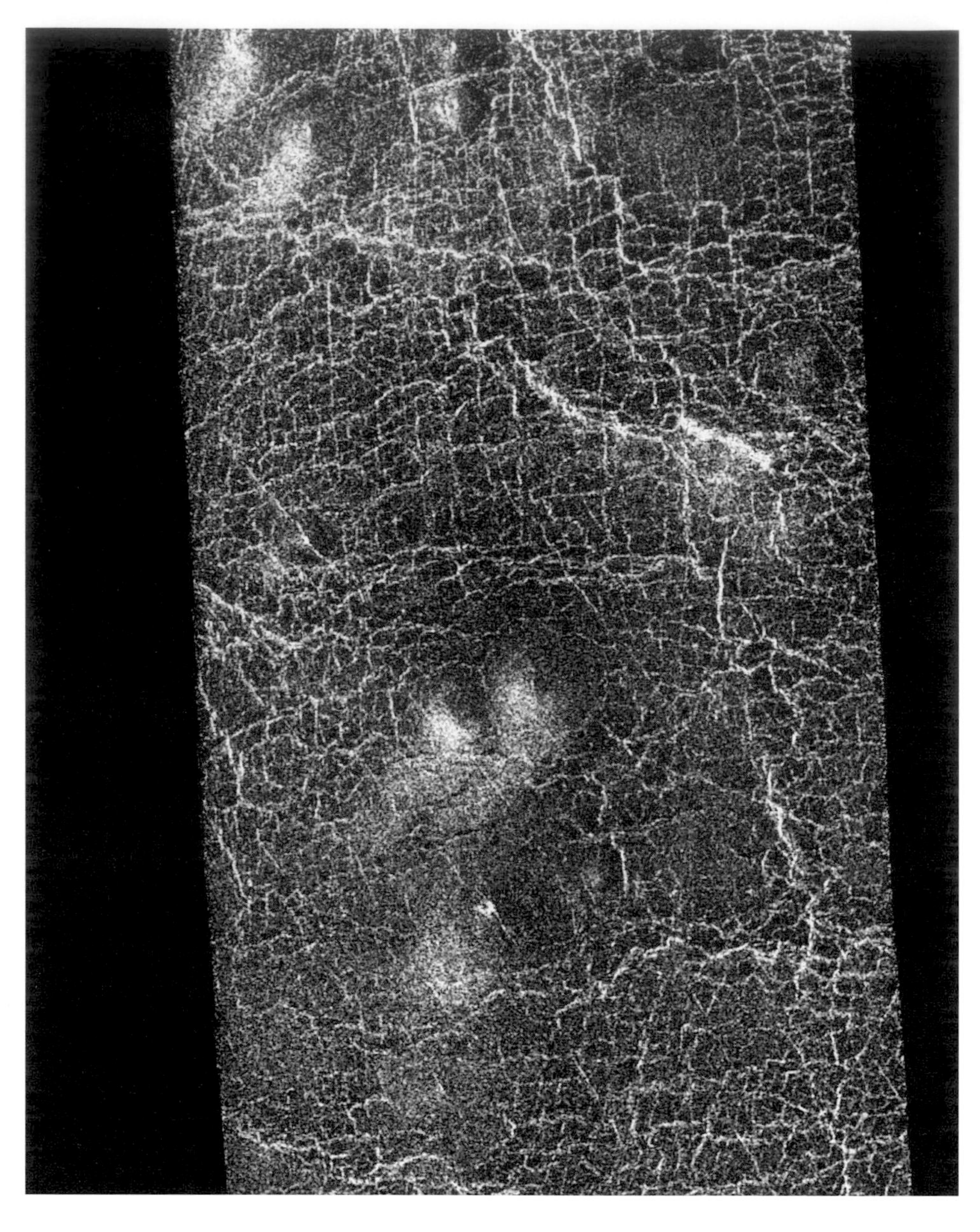

numbers on the rolling plains but be scarce on the lowlands. However, this difference in distribution may be more apparent than real, since it is possible that any coronae which formed on the lowland surfaces, subsequently were inundated by flood lavas.

Geography of Venusian plains

To the south of Ishtar Terra are the extensive lowlands of Sedna and Guinevere Planitiae whose deepest levels descend 1 km below datum. These plains are punctuated by the isolated highland massifs of Eistla, Alpha and Bell Regiones. On the plains surface are large numbers of coronae, the peculiarly Venusian deformational structures apparently formed above mantle plumes or hot spots. East from Guinevere Planitia the surface rises along a belt of equatorial highland massifs which together comprise Aphrodite Terra, the most extensive highland region on the planet. This is composed of a number of individual tessera massifs and volcanic rises. The deep rifts, Diana, Dali and Artemis Chasmata, separate these from one another.

To the north of Aphrodite are the plains of Leda, Niobe and Rusalka Planitiae, while to the south lie the extensive plains of Aino Planitia. Eastwards from Aphrodite a further series of rifts extend towards the N/S-trending highland massifs of Beta–Phoebe–Themis Regio, massive volcanic complexes which have grown along a meridional system of major rift faults. Beta is connected to Phoebe Regio by the 2500-km-long, 90-km-wide rift of Devana Chasma which, at its deepest point, descends 2.5 km below datum.

The northern polar region is a broad plain crossed by tectonized zones and surrounded by highland massifs. The ridge-belts which traverse it are several hundreds of kilometres in length and tens of kilometres in width. The longest continuous belt runs along the 200° meridian from Atalanta Planitia into the polar region and then turns south along 80°E, where it abuts against Ishtar Terra.

Radar-dark plains adjacent to the southeast margin of Artemis Chasma. Note the curvilinear wrinkle ridges and the lobate volcanic flows with differing radar signatures. [Framewidth 530 km.]

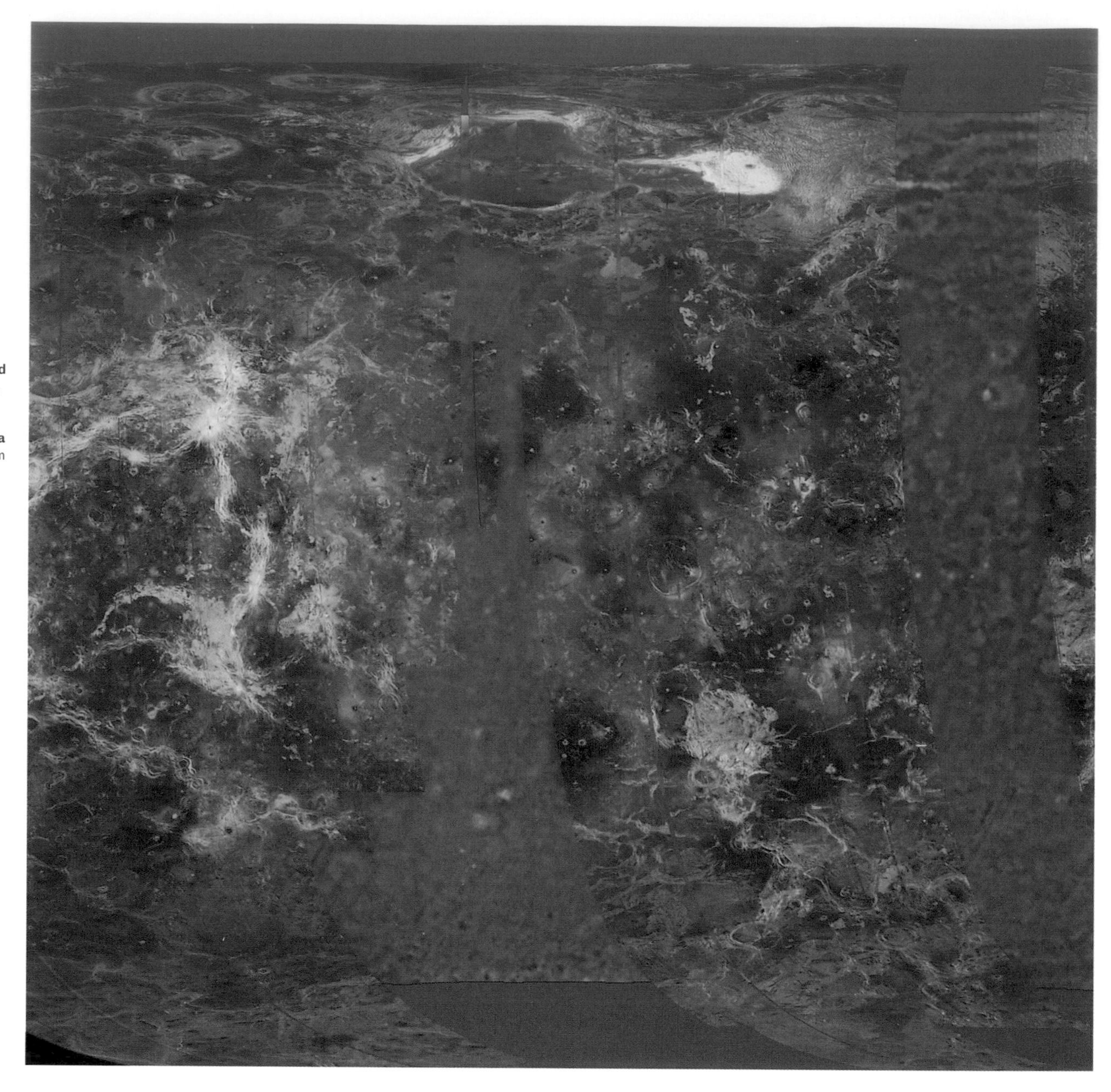

The western half of Venus is shown in this Magellan mosaic. It clearly indicates how the plains are punctuated by highland massifs, such as Ishtar Terra (top), **Beta–Phoebe Regio** (left) **and Alpha Regio** (towards bottom right).

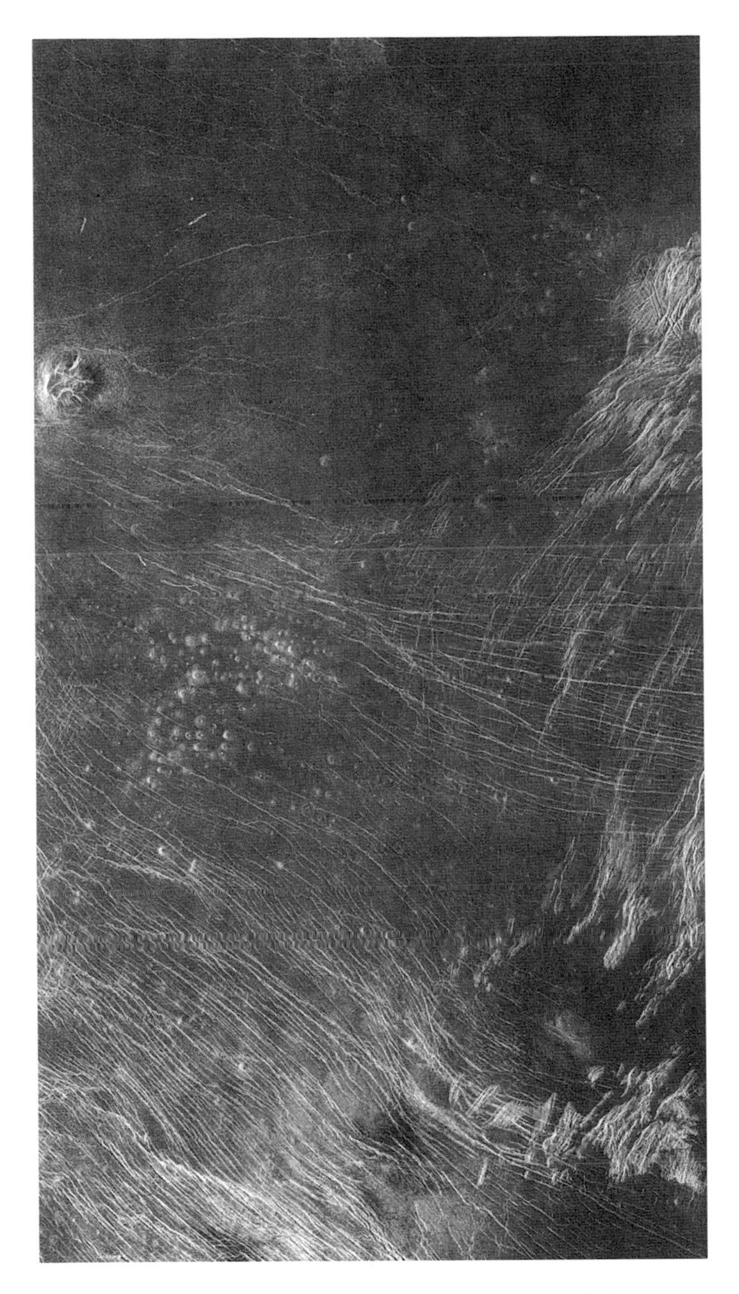

Seen in this image is a typical Venusian shield field, a cluster of small volcanoes between 1 and 5 km in diameter. To the north is a steep-sided, fractured dome – nicknamed a "tick" by NASA scientists – which appears to indicate somewhat more viscous magmas were able to arch up the Venusian crust.

Some of the southern polar region still remains to be imaged.

Details of the plains units

The fine detail on the plains' surfaces was gradually unveiled by the Magellan radar mapper. The highly complex volcanic and tectonic history of coronae and related features also became apparent. In the initial Magellan mission summary, four major plains types were recognized:

(i) Smooth plains which, as the name implies, are relatively featureless, have no discernible volcanic flow features, few linear structures, and tend to have dark radar signatures (but range up to moderately radar-bright). Such plains have their origin in volcanic flooding, presumably by very fluid lavas or by the coalescence of individual low shields.

(ii) Reticulate plains are distinguished by having one or more sets of somewhat sinuous, radar-bright lineaments. Their morphology suggests an origin in volcanic flows or low shields which either have embayed older units, or have been tectonically deformed.

(iii) Gridded plains have intersecting orthogonal sets of radar-bright lineaments, regularly spaced, which extend for hundreds of kilometres. The spacing of these features tends to be closer than those in (ii) and typically is less than 5 km. Complex deformation is a characteristic.

(iv) Lobate plains comprise overlapping lobate volcanic flows with variable radar signatures which extend for tens to hundreds of kilometres. Such plains are traversed by few if any linear structures.

Most of the different types of plain are traversed by wrinkle ridges which show up as radar-bright lines. As a rule, individual ridges are located between a few kilometres

(a)

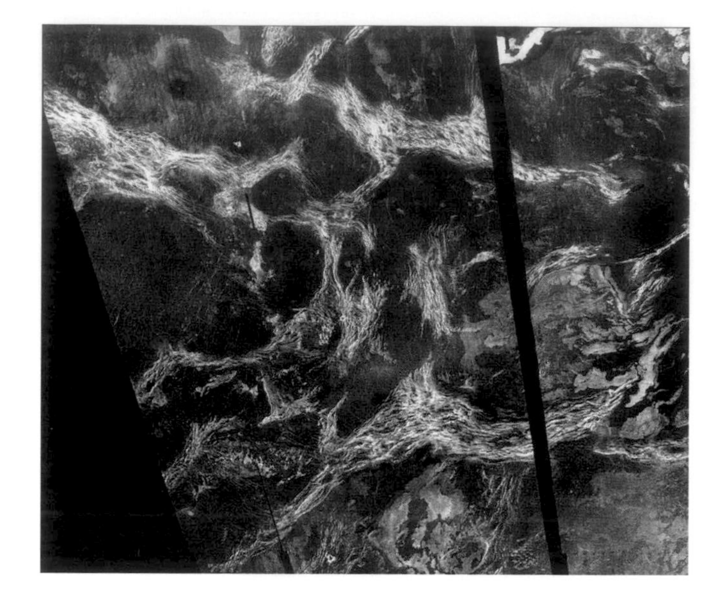

(a) Radar-bright ridge- and fracture-belts cross the plains of Lavinia Planitia. Note also the intermediate-brightness lobate flow features and small tessera massifs. [Framewidth 1843 km.]

(b) Sketch map of northern hemisphere of Venus showing ridge-belt fan extending from the north pole towards Aphrodite Terra.

(b)

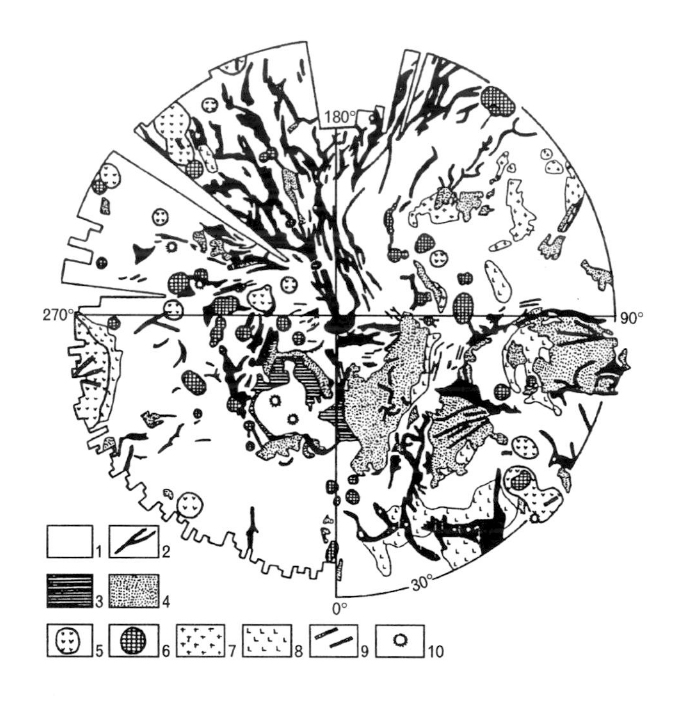

Tectonic map of the North part of Venus. 1 – plains, 2 –ridge belts, 3 – mountain-ridge belts, 4 – tesserae, 5 – dome-like uplifts, 6 – coronae, 7 – hummocky plains (probably tesserae partly buried by plain materials), 8 – diapiric plains, 9 – rift depressions, 10 – shield volcanoes.

to 20 km apart. Similar structures abound on the plains units of both the Moon and Mars, where the ridges show quite strong preferred orientations within different regions (as they do on Venus). Ridge formation – presumably by compressional forces – was, and may still be, an ongoing process connected to local stress fields that operated while the plains units were being emplaced.

Radar-bright, narrow grooves also extend for anything between 25 and 200 km across the Venusian surface, with spacings of between 30 to 100 km. These are extensional features and most appear to be graben (trough faults). As with the ridges, they tend to share a common strike over quite large areas but always lie normal to the ridges, which gives the plains of regions like Lavinia and Guinevere Planitiae a distinctive orthogonal imprint.

Ridge- and fracture-belts

Characteristically developed within the Venusian plains are linear deformational features termed ridge-belts. These comprise narrow ridges and arches which rise a few hundred metres above the surrounding plains, are up to 300 km wide, and may extend for hundreds to thousands of kilometres as radar-bright swaths across the planet's surface. They show up dramatically in Arecibo radar mosaics. The zones of deformation which ridge-belts represent are not distributed uniformly within the plains areas, tending to be concentrated in or near major lowlands, e.g. Lavinia, Atalanta and Vinmara Planitiae. Other deformation belts are located southeast of Alpha Regio and south of Aino Planitia, in the southern hemisphere, and north of Bereghinya Planitia, in northern latitudes. The very broad spatial distribution of these features has important implications for the stress state within the planet.

Types of belt

The belts have two regional manifestations: first, parallel or subparallel networks and fan-shaped patterns within lowlands like Lavinia and Atalanta Planitiae, and second, as more nearly orthogonal networks in close proximity to raised tessera blocks. The first type of ridge-belt is the one which shows up on Veneras 15 and 16 images of the northern hemisphere. This type is generally sufficiently broad and has sufficient vertical relief to show up on most Magellan altimetric data. The second type typically shows little or no relief, yet the individual narrow ridges are evident whatever the incidence angle of the radar beam. To facilitate distinguishing between the two classes, it is customary to term the first type fracture-belts and the second, ridge-belts. Frequently the two classes of belt are found together, either in combination as a single complex belt, or as distinct belts with different strikes. Within such composite belts the ridges may share the overall strike of the fracture-belt or may strike at some (often markedly different) angle to it.

The fracture belts found in Lavinia Planitia have been very closely studied. They have been found to comprise families of grooves which range in width from a few kilometres down to the limit of Magellan's

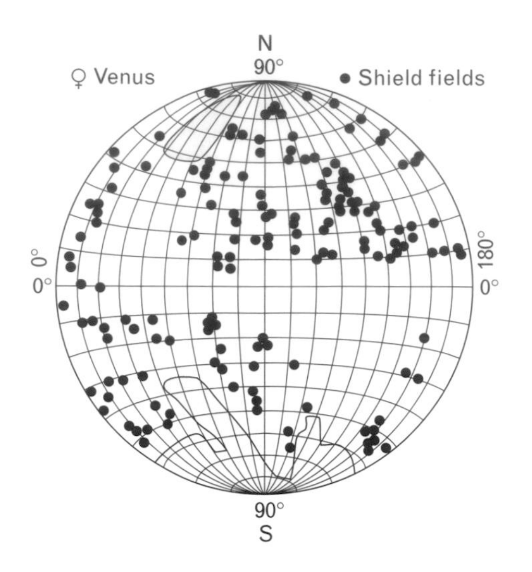

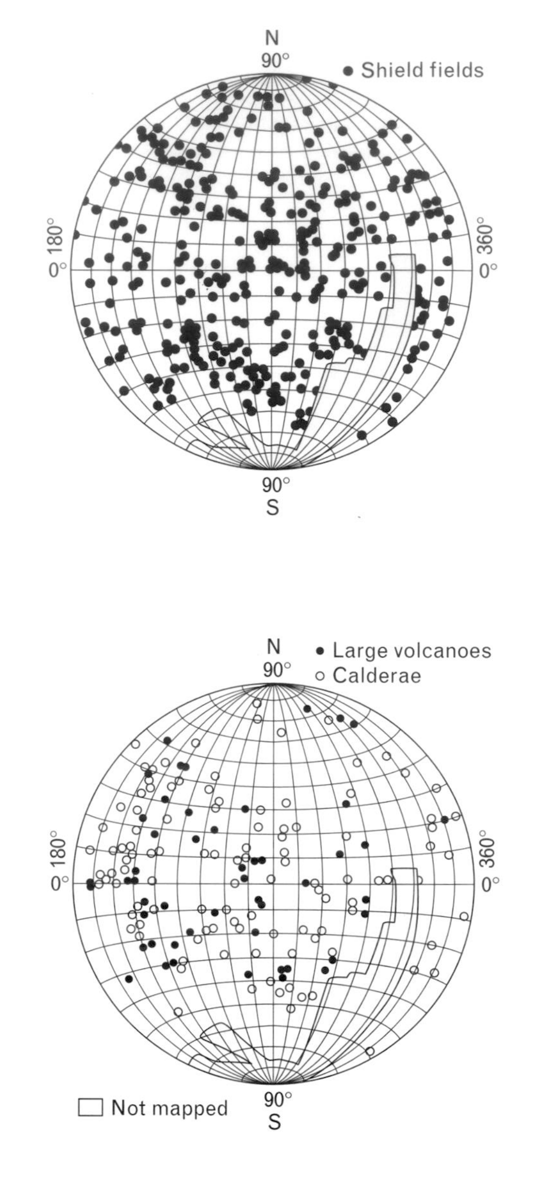

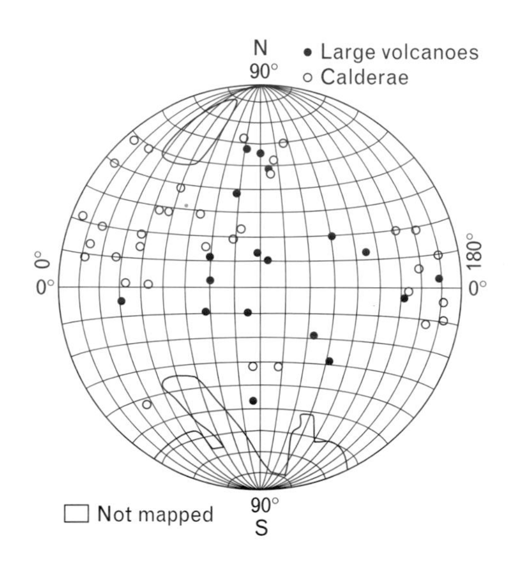

Distribution of shield fields and large volcanoes/calderae over the surface of Venus.

resolution. That they were formed over either a lengthy period of progressive deformation, or during repeated phases of deformation, is indicated by their typical winding courses and cross-cutting relationships. In some places the spacing of the fracturing is very close indeed, with the intensely-fractured bands being separated by relatively little-deformed rocks. One other interesting observation in connection with the fracture-belts is that the prominent faulting which is their characteristic is extremely similar to the less strongly developed fracturing in the adjacent plains. There must be a close genetic connection between the faulting typical of fracture-belts and the production of grooves on the Venusian plains. Most of the tectonic features found within the plains, including the ridge- and fracture-belts, demand only modest strains and horizontal displacements of the order of tens to hundreds of kilometres, assuming crustal thicknesses of between 20 and 30 km.

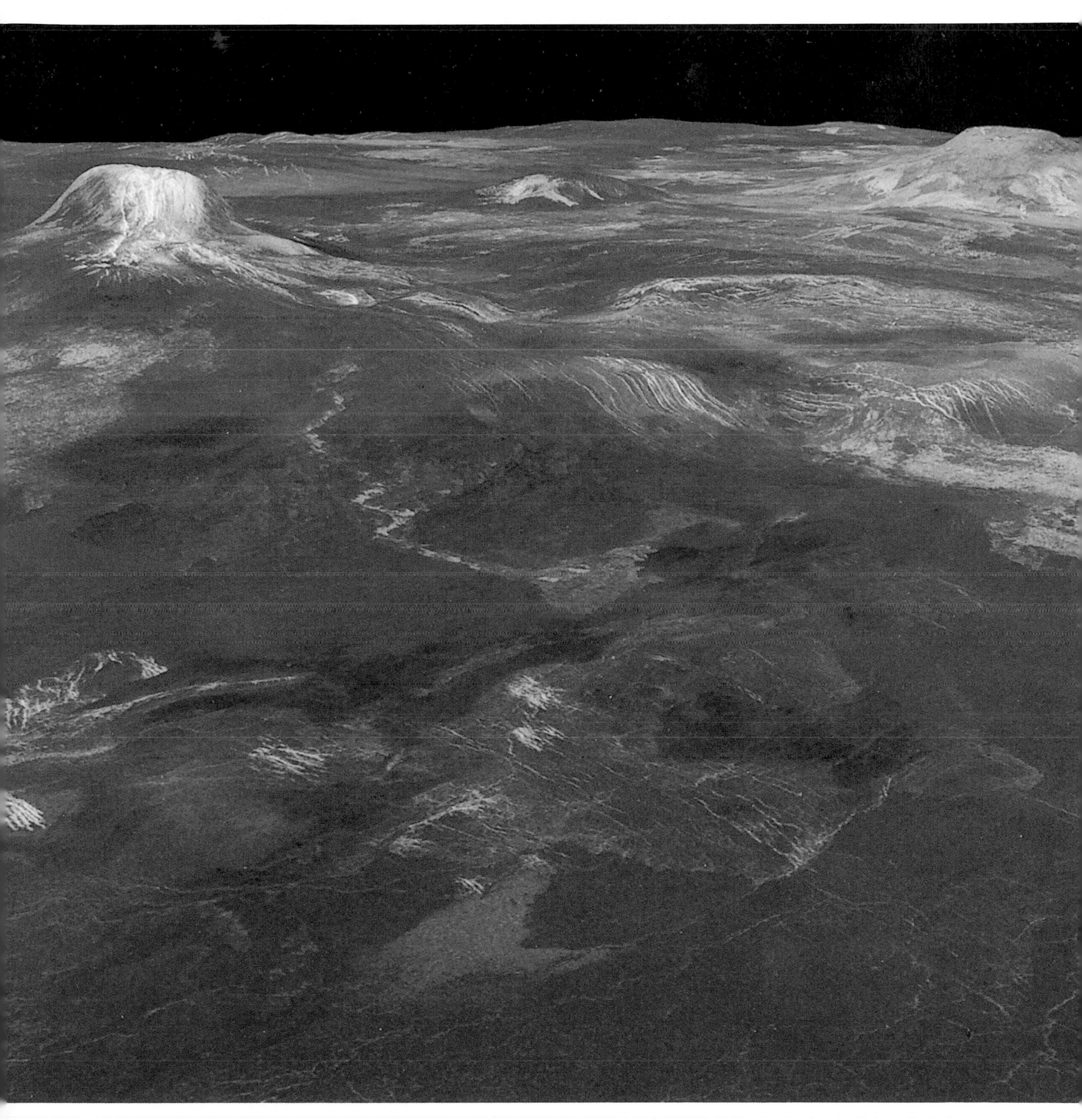

Perspective view of the large shield volcanoes, Sif and Gula Mons, with their associated lava flows. Gula Mons, located just below the horizon on the left of the image, is 3 km high and has a diameter of approximately 250 km. Sif Mons – towards the right – rises 2 km above the surrounding plains, and has a basal diameter of 300 km.

A segment of an anastomosing lava channel crossing the ridged and fractured plains of Atalanta Planitia. The entire channel is about 2 km wide, and is 7000 km long, that is, considerably longer than the Earth's Nile river system! Such features are a manifestation of the thermal erosion of the plains rocks by highly fluid, hot lava.

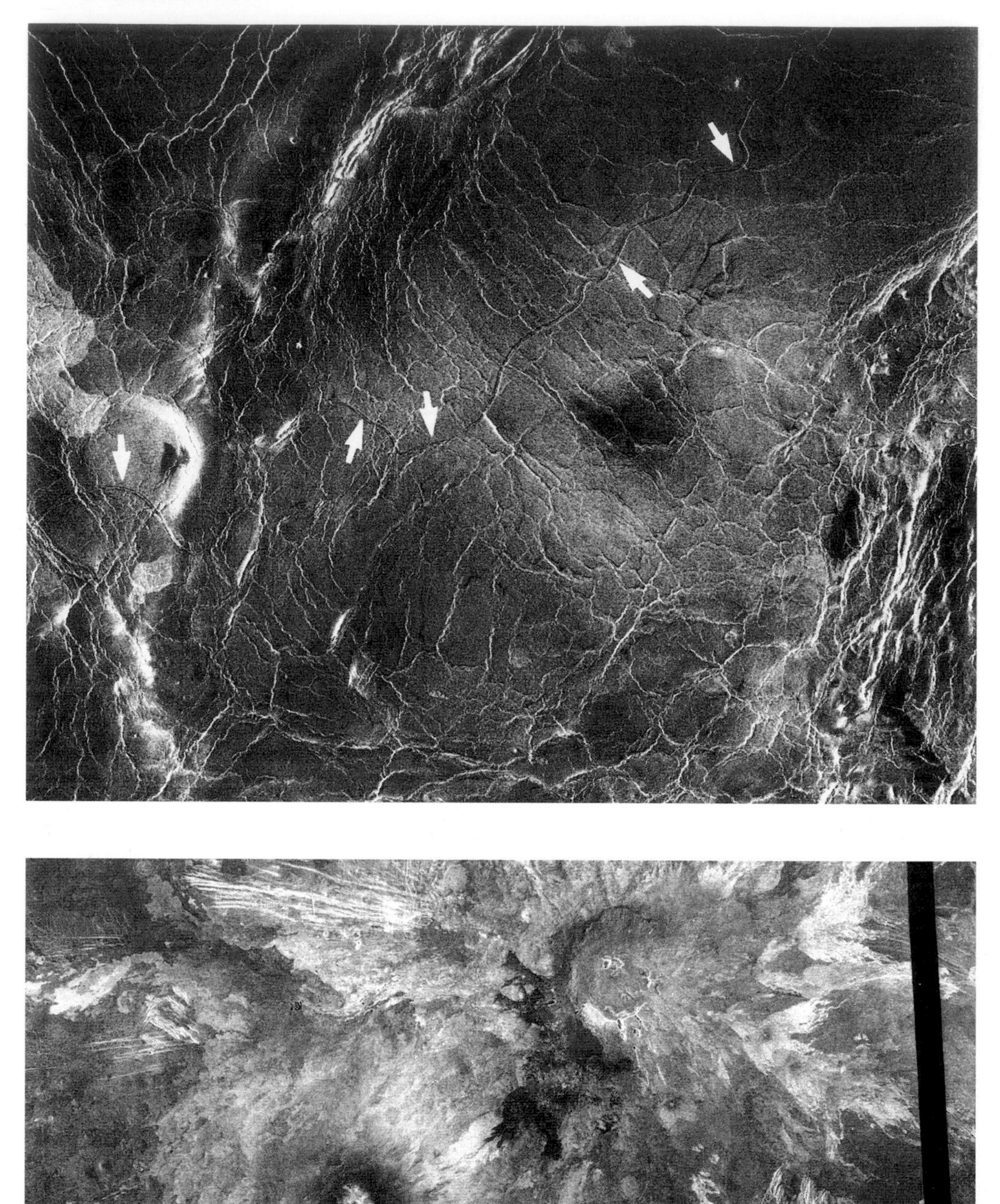

A series of massive volcanic flow fields surround the central region of the shield volcano, Sif Mons, located in western Eistla Regio. The complex superposition of flows with differing radar signatures shows how the structure was built up over a lengthy period of effusive volcanism.

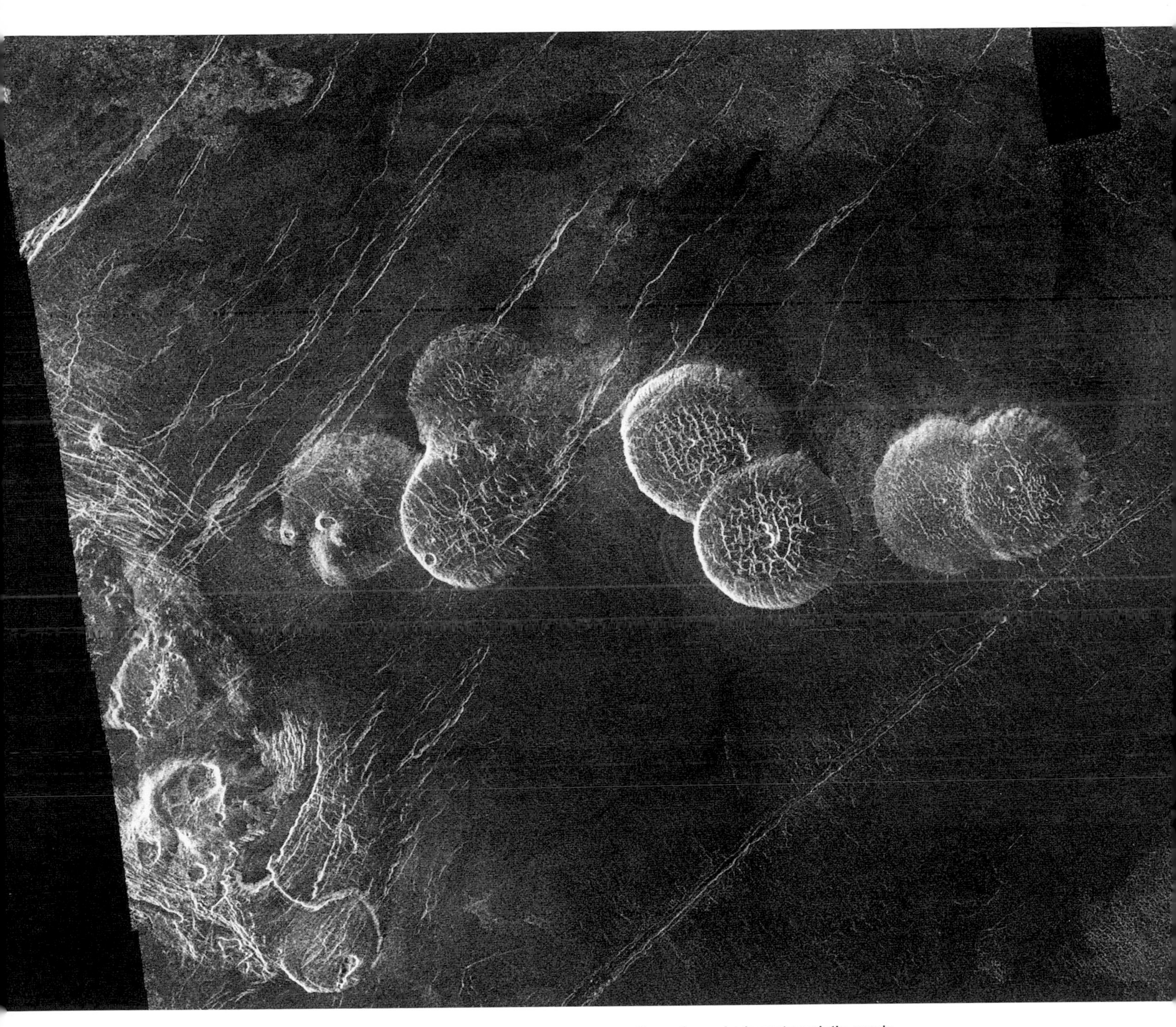

Group of pancake domes towards the margin of Alpha Regio. The seven circular, flat-topped volcanic excrescences average about 20 km in diameter and 750 metres in height. They have strongly-fractured carapaces and summit depressions.

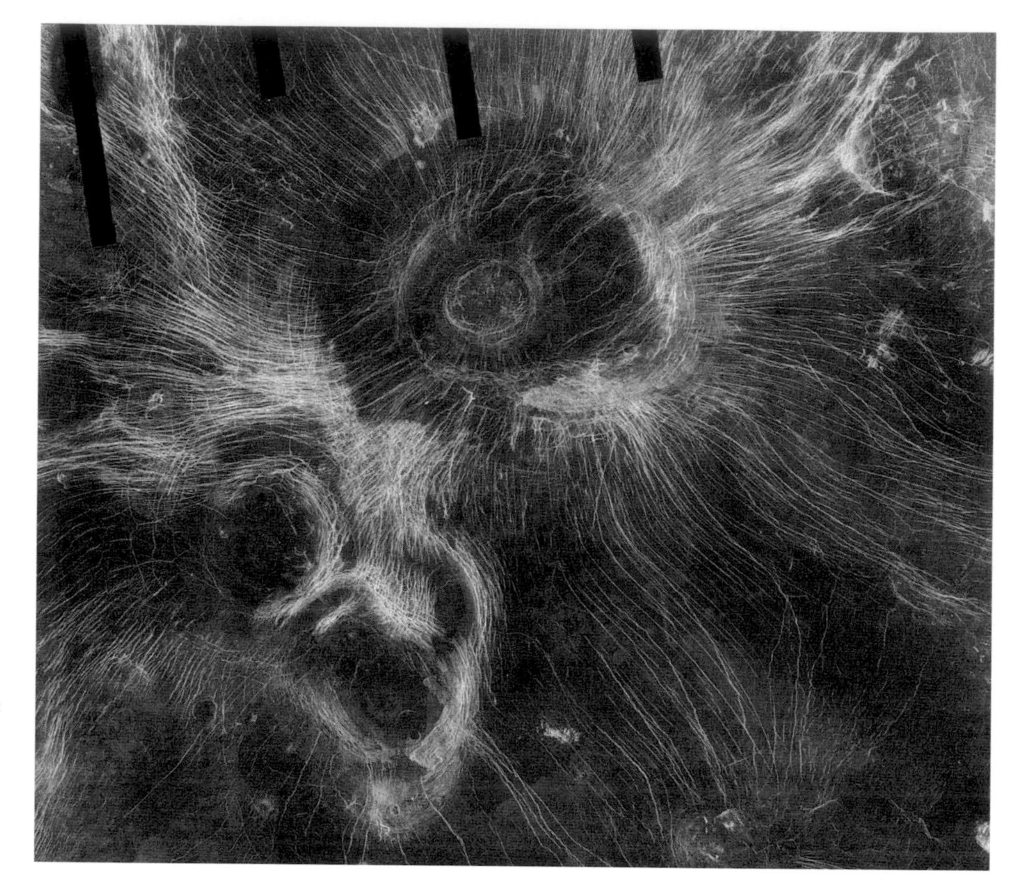

Multiple corona structures in Themis Regio. The larger structure measures 380 km × 200 km and is linked to two further structures. Note the prominent radial and concentric fracturing, typically associated with these uniquely Venusian volcano–tectonic structures.

Volcanic features of the plains

There is little doubt that volcanism has been responsible for generation of the Venusian plains. The kinds of plains-forming volcanic structures developed – long flow-fields, caldera depressions, domes, low shields – are typical of plains-forming basaltic volcanism on the Earth and other planets, and a basaltic composition for Venusian magmas is implied by the Venera and Vega geochemical data. In addition to the smaller-scale features there are numerous larger volcanic centres, first revealed clearly by Venera 15/16. We now know there are 156 large volcanoes over 100 km across, 274 volcanoes in the range 20–100 km, 86 caldera-like landforms between 60 and 80 km diameter which are

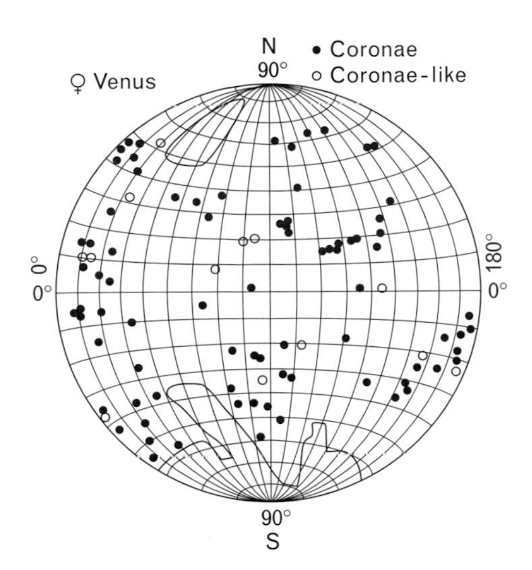

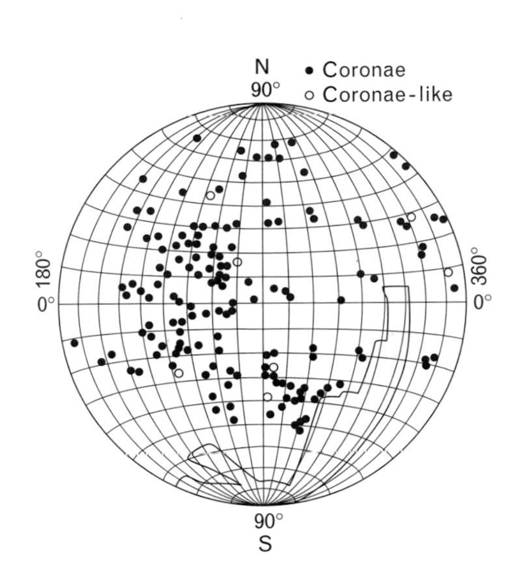

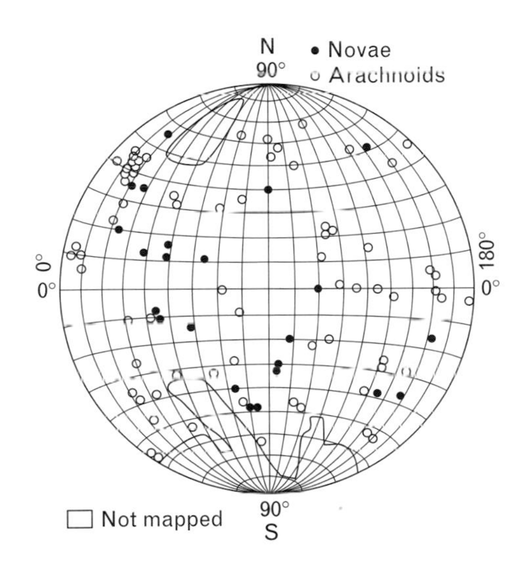

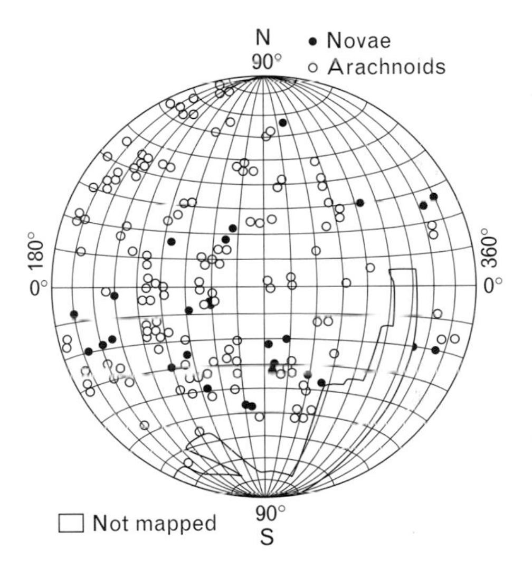

Distribution of coronae, novae and arachnoids on the surface of Venus.

not associated with large shields, 175 annular concentrations of fractures and ridges termed coronae, 259 arachnoids and 50 novae (foci comprising radial fractures forming stellate patterns). In addition there are 550 clusters of small <20 km volcanoes called shield-fields. This is the tally of a planet with a long and intensive volcanic history.

The global distribution of volcanic landforms contrasts strongly with the strongly-focussed pattern of terrestrial volcanism along lithospheric plate boundaries. The absence of chains of large shield volcanoes on Venus precludes the possibility of large-scale crustal movement above static hot spots. However, the distribution of volcanic landforms is not totally random: more than

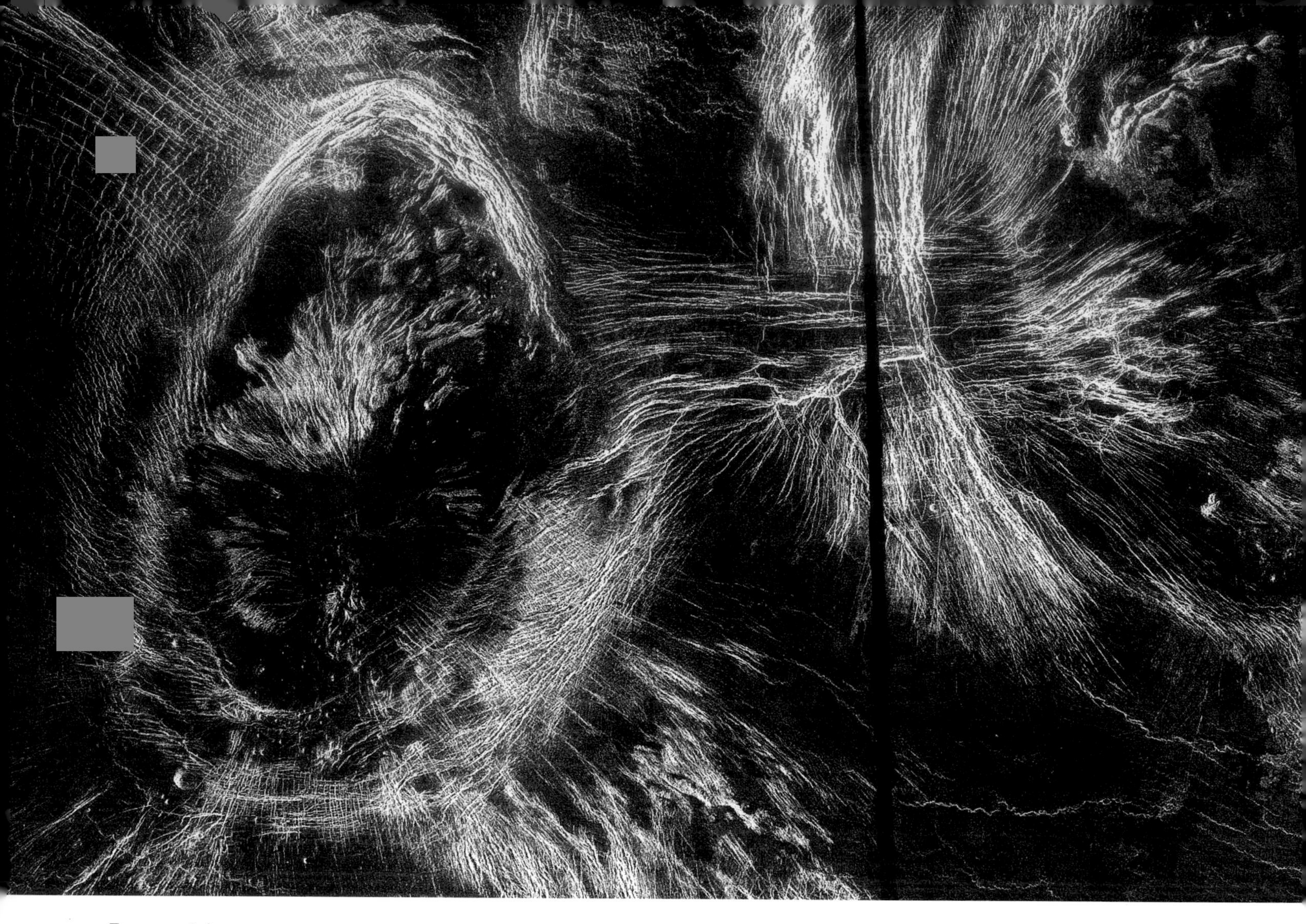

Two connected coronae in the Fortuna region of Venus. That on the left, Ba' het, is roughly 230 km across and is joined by a complex of fractures to Onatah Corona, over 350 km in diameter. Both structures are surrounded by a ring of ridges and troughs and contain, at their centres, volcanic flows and domes.

70% of all shield-fields, coronae and large volcanoes lie within one-half of the surface. Furthermore, there is a definite deficiency of constructional landforms in the lowlands, indicating the likelihood of extensive fissure-related volcanic flooding of these regions. Major lava channel systems also flow down towards the lower-lying areas.

Most Venusian volcanoes have relatively modest slopes; however, distinctive pancake domes have flat tops and steep sides which suggest the rise of lava more viscous than basalt. Individual volcanoes share many of the characteristics of terrestrial shields or domes, but there are no terrestrial analogs of Venusian coronae, novae and arachnoids. These structures are highly fractured, have generally raised interiors, depressed surrounding moats and associated volcanic flow features, small shields or domes. They range in size from 50 km to over 2000 km (Artemis Chasma) and, together with the large-scale volcanic rises of Venus, are believed to be located above rising mantle plumes or "hot spots". Coronae may occur in chains, such as that associated with Parga Chasma, but generally there is no well-defined linear distribution. The escape of heat from inside Venus evidently did not take place along lithospheric plate boundaries, at least not during relatively recent times.

THE VENUSIAN PLAINS – II

8

The Venusian plains play host to many impact craters. The impact record and the style of cratering and ejecta deposition is, however, different from that on airless worlds. While Venusian winds are of low velocity, they have been responsible for producing wind streaks and fields of dunes in some areas.

While the plains of Venus owe their origin to volcanic processes, radar images also show many impact craters with which are associated ejecta deposits unlike those seen on other worlds. The fragmental material generated by impacts has, in places, been redistributed by the torpid Venusian winds, forming dunes and haloes around some craters. Analysis of Pioneer-Venus reflectivity data suggests that not more than 5% of the surface is covered in *regolith* more than several tens of centimetres thick, however over the majority of Venus, bare rock surfaces tend to dominate. Interestingly, within the regions of tessera terrain there are abundant rock fragments which originate from a hitherto unknown but widespread mass-wasting process.

The present lack of water on the surface of Venus, naturally means there cannot be river systems, oceans, lakes or ice sheets. However, that some form of surface modification has occurred on Venus was clear from Veneras 15 and 16 and Earth-based images of several fresh craters, which are surrounded by prominent radar-bright haloes. About 25% of all craters imaged by Veneras 15 and 16 have these.

Winds on Venus

What of the wind velocities near the ground on Venus? Anemometers at Veneras 9, 10 and 13 sites registered winds speeds of between 0.3 and 1.0 m s^{-1}. Furthermore sequential TV pictures at the last site demonstrated that a clod of regolith a few centimetres across, which had been churned up during the landing of the probe, was entirely removed during the hour long interval during which observations were made.

The wind velocities recorded theoretically are capable of mobilizing at least fine sand and silt-grade particles. However, some more precise data were required and these were provided by carefully-controlled wind tunnel experiments. These proved highly informative, revealing that within a pure CO_2 atmosphere, 0.6 km s^{-1} winds will entrain quartz grains 100μm in diameter, while a wind of velocity 3 km s^{-1} will shift 1 cm particles. Even by the time Magellan had completed but half of its initial 8-month-long mapping cycle, it had revealed a number of different kinds of surficial deposits not previously identified

Wind tunnel experiments showed that the most easily moved particles have a size of around 75 μm, under simulated Venusian conditions. They have demonstrated also that *microdunes* form in fine-grained quartz-sands (50–200 μm diameter) under appropriate conditions; however, if wind velocities exceed about 1.5 m s^{-1}, dunes apparently are suppressed and resultant bedforms are flat and featureless. The implication here is that if relatively short periods of enhanced wind speed buffet existing dunes, these may be destroyed – a phenomenon which could account for the apparent scarcity of such structures in pre-Magellan imagery. The generation of microdunes, as opposed to other bedform structures, i.e. surface ripples, waves and featureless surfaces (all produced during the experiments), is dependent primarily on wind speed.

The experiments also showed that Venusian aeolian transport differs significantly from that on Earth, particularly with respect to the rolling mode of sediment transport. This can be maintained for long periods under Venusian conditions without leading to the cascading effect which, on Earth, will suddenly transform the surface of a bed into a saltation cloud. To some extent, therefore, aeolian environments on Venus share some of the characteristics of aqueous conditions on the Earth.

on Venus; several of these are associated with impact craters. Hummocky materials with high radar backscatter outcrop both within and surrounding many craters; these are presumed to have an impact origin. Also having bright radar signatures are lobate materials whose distribution appears to be topographically controlled. These have been interpreted as volcanically-generated *outflow deposits*, the volcanism having its origin in impact.

Wind streaks

The wind velocity measured at the Veneras 9 and 10 sites ranged between 0.5 and 1.0 m s^{-1}. Such velocities coupled to the dense Venusian atmosphere are quite capable of mobilizing fine sand or silt. Not surprisingly, therefore, Magellan revealed a small number of major dune fields and various types of wind streak. Wind streaks – well known from both Mars and the Earth – are the more common, over 3000 having been catalogued. While the greatest incidence of streaks is in the equatorial plains, wind streaks are found at most latitudes and elevations. This demonstrates that wind activity is widespread on Venus.

Wind streaks are well known from Mars, where they range in size from a few centimetres long to 115 km. They represent the prevailing wind direction at the time and place of their inception and have been used to establish details of local, regional and global atmospheric circulation. They occur in a variety of forms which may show radar-dark, radar-bright or radar-mixed reflectivity against the background on which they have developed. On Mars, dark wind streaks are found where bedrock has been swept clear of loose sediment, or where lag deposits of

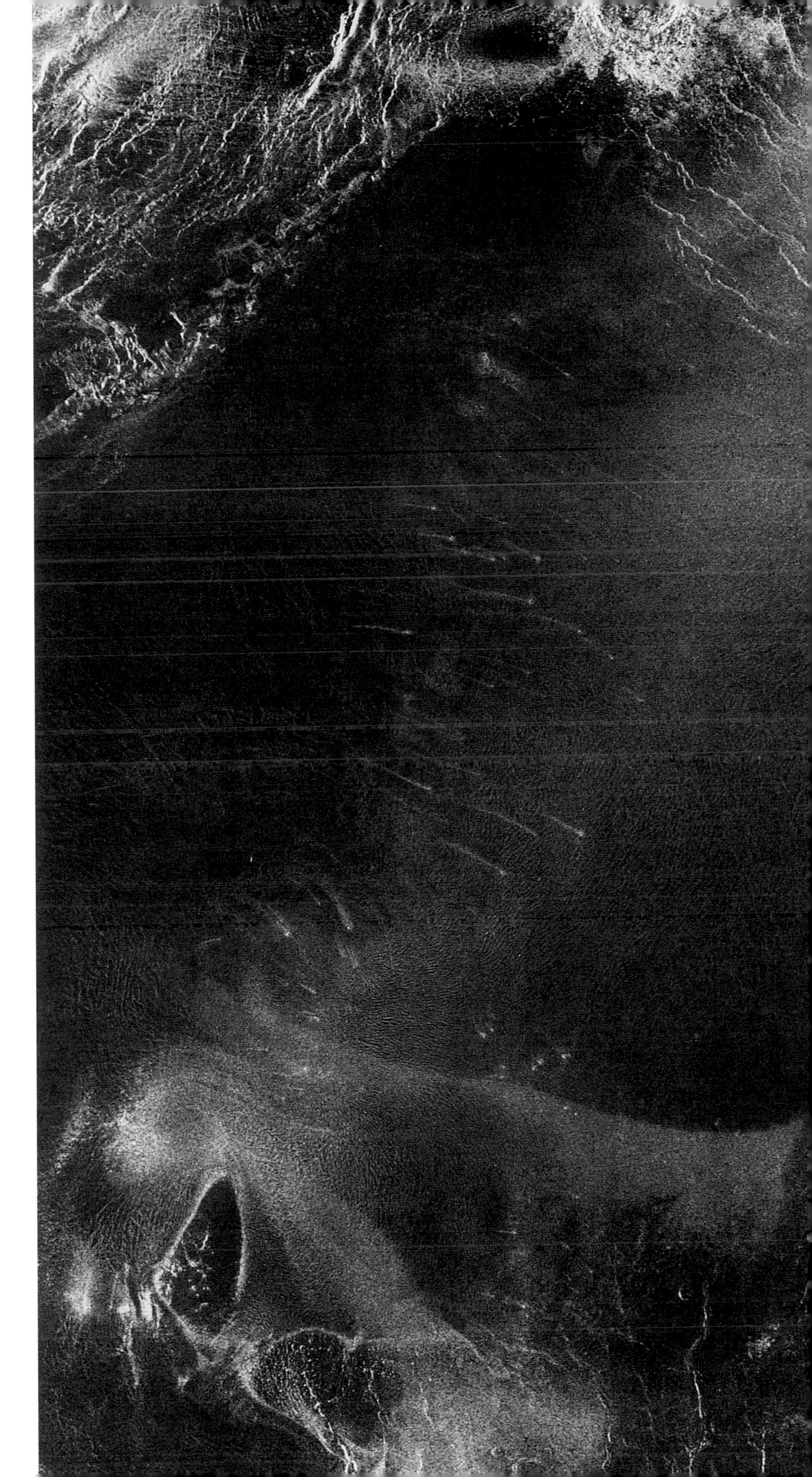

Radar-bright wind streaks and elongate dunes are revealed to lie in the broad valley between Ishtar Terra and Meshkenet Tessera. Wind activity on Venus appears most closely associated with impact crater ejecta and certain parts of tessera blocks. The mosaic is 190 km wide and 340 km long.

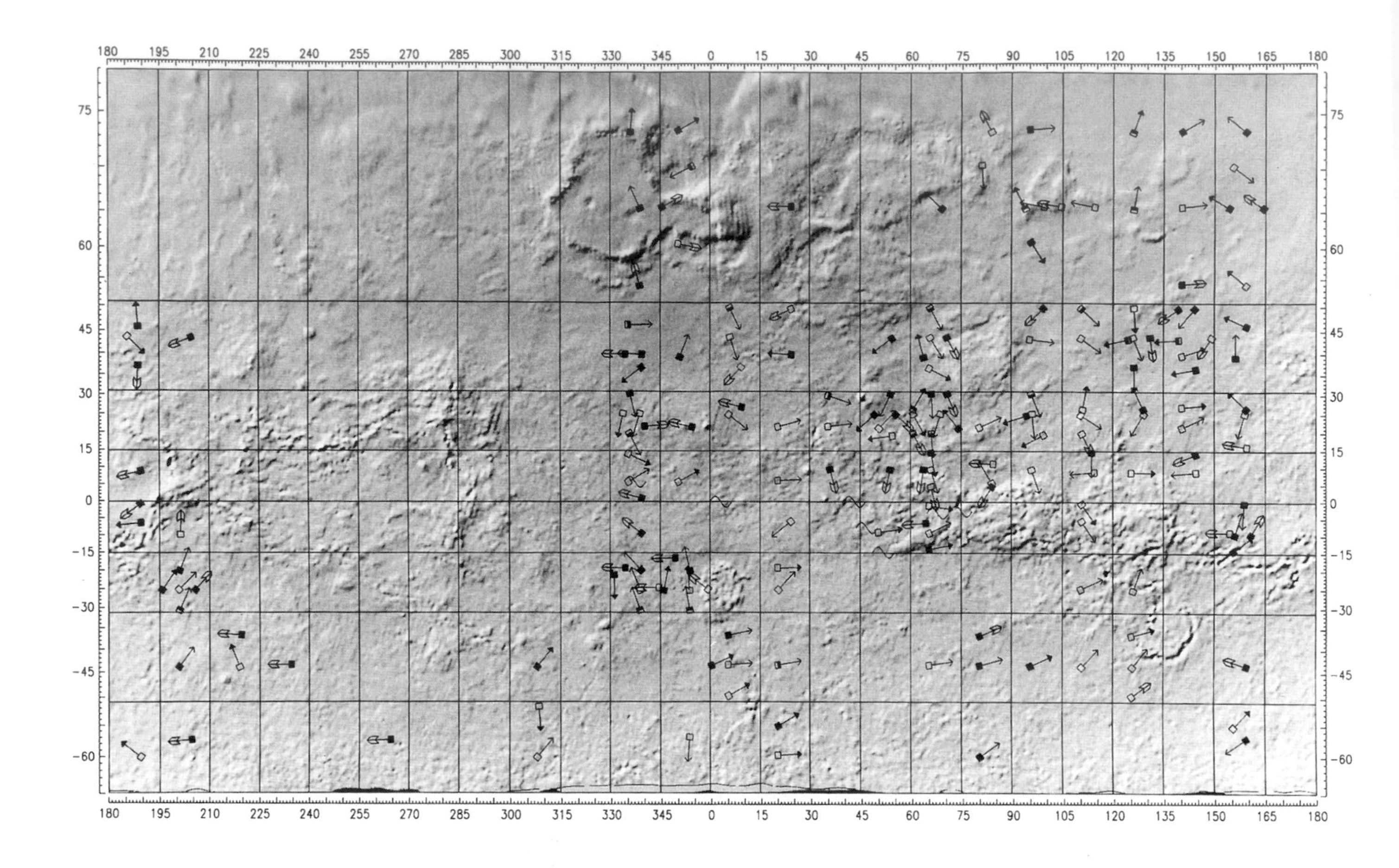

Distribution of wind streaks over the surface of Venus. (After Ronald Greeley and colleagues.) Downwind direction is shown by head of arrows.

Streak type

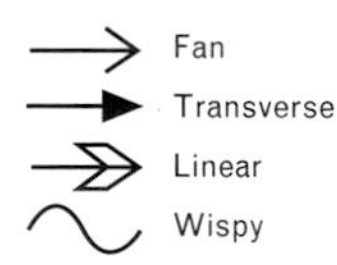

Reflectivity

☐ Bright
■ Dark
◧ Mixed

coarse grains have been modified by the removal of smaller, more reflective particles by deflation. In contrast, bright streaks are attributed to dust deposition in the lee of obstacles, probably under conditions of atmospheric stability.

Particularly prominent streak-like features are associated with Sif Mons and the impact crater, Carson, situated west of Alpha Regio. These form immense haloes. Typically the vertices of such features are located on the eastern edge of a paraboloid-shaped region, with the long axis striking N/S. East-west dimensions of between 500 and 1000 km are typical. The low emissivity values and low backscatter recorded in these regions imply that they are smoother than their immediate surroundings.

Dunes

One of the first reports of a possible dune-field was associated with a 65-km-diameter impact crater located west of Alpha Regio, near the crater Carson, and 100 km north of the crater Aglaonice. The diagnosis was largely based on the speckle-type radar return which this region gave and which terrestrial dunes are known to give. Its existence was confirmed by Magellan as it mapped the planet. The Aglaonice dune-field is characterized by dunes 0.5 to 5 km in length and covers an area of approximately 1290 km^2. It is associated with radar-bright outflow deposits associated with the impact crater from which it derives its informal name. This outflow material evidently has been reworked by aeolian agents to produce both

Linear streaks associated with Mead crater are believed to represent yardangs.

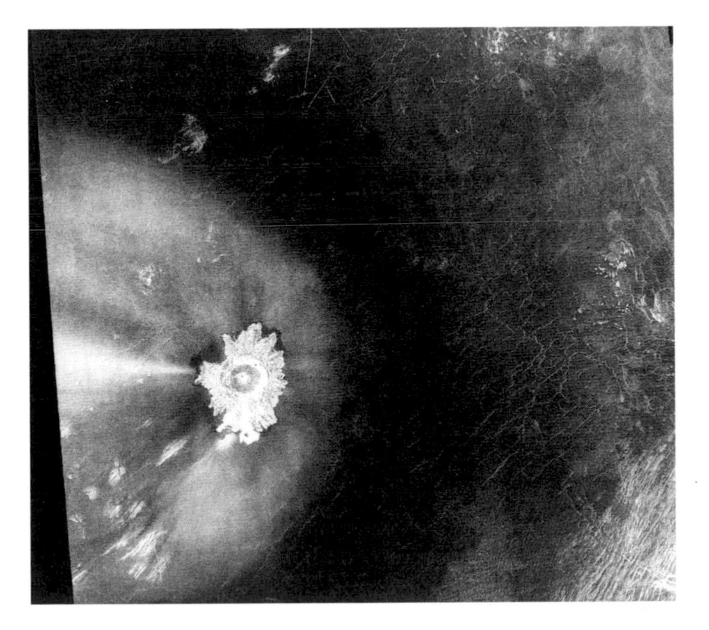

Wind streak associated with the impact crater, Adivar.

Dark halo believed to be produced by "atmospheric cratering" of the surface. It is envisaged that potential impactors, too small to reach the surface, break up in the Venusian air, but that the atmospheric shock wave they engender is sufficient to blast the surface below the impact point clear of surface debris. The halo is about 20 km across.

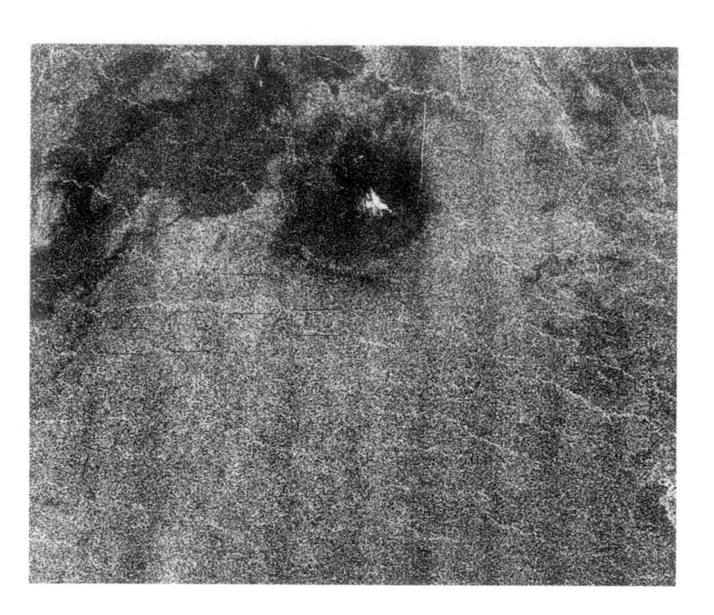

Streaks occur in a variety of forms which may show radar-dark, radar-bright or radar-mixed reflectivity against the background on which they have developed. A number of basic outcrop patterns can be identified, including fan-shaped, wispy, linear, transverse-ragged and transverse-smooth streaks. Fan-shaped streaks characteristically are associated with positive landscape features such as small hills or domes; most have a radar-bright signature. Linear streaks, in contrast, are radar-dark and typically are at least 10 times longer than they are wide, occurring in groups of at least six individuals. Wispy streaks meander cross-country, their width changing over quite small distances; they are often located adjacent to ridges with a parallel strike. Some groups of wispy streaks apparently are associated with impact craters, in which occurrence the streaks occur in groups that give the impression of being arranged approximately radial to the focal crater. However, because of their rather irregular nature, it is not usually possible to discern their point of origin with any real degree of certainty. Transverse streaks also occur in groups, often associated with fractures or ridges; these may be either radar-dark or radar-bright, and generally are oriented normal to the inferred prevailing wind direction.

Global analysis shows that the streaks are found at all altitudes, and tend to be orientated downwind and toward the equator. Most are randomly oriented with respect to slope, and located where the gradient is less than 2°. This is consistent with surface winds related to a Hadley circulation in the lower atmosphere. In such a pattern the circulation redistributes solar energy absorbed in the lower levels of the atmosphere and near the ground near the equator. Such a meridional circulation is symmetrical about the planet's equator and involves surface winds blowing towards the equator, upflow over the equator itself, poleward winds aloft, and air downflow at high latitudes.

the transverse dunes and wind streaks. The orientation of the dunes, which are presumed to be transverse types, together with several bright wind streaks in the same region, suggest that the prevailing winds currently blow towards the west.

A second such field is located much further north, in a valley between Ishtar Terra and Meshkenet Tessera. This has been called the Fortuna-Meshkenet dune-field, and it covers an area of approximately 17,120 km^2. Individual dunes are between 0.5 and 10 km long, are around 200 m wide, and have an average separation of around 0.5 km. As with the Aglaonice field, this second field is associated with numerous prominent wind streaks. In the southern part of the field, the dunes and streaks indicate a flow of wind from the southeast, while towards the north the orientations of the aeolian landforms suggests a westward flow. The radar-bright wind streaks located within the field apparently have their origin in several small radar-bright cones, whose similarity in radar return suggests the cone material is their source.

The area surrounding the impact crater Mead, situated on the northwestern flank of Aphrodite Terra, hosts the greatest concentration of aeolian features on Venus. These include the only occurrence of potential Venusian yardangs. Terrestrial yardangs are narrow, often undercut, ridges of rock, separated by long corridors which have been eroded by wind-blown sand; they are roughly parallel and have their long axes parallel to the prevailing wind. The features believed to be Venusian analogs of this landform type comprise two sets which together include one hundred individuals; the orientation of both sets indicates a prevailing northeast to southwest wind flow. Since most terrestrial yardangs develop in relatively friable and often well-jointed materials (e.g. lacustrine sediments, volcanic ashes), it seems likely that the deposits associated with the formation of Mead crater have provided suitable easily-erodable materials for yardang formation and that winds in this area have remained constant in direction at least in the relatively recent past.

Deposits associated with impact craters

Impact craters also exist in abundance on the Venusian plains, although their numbers are smaller than those found on the ancient surfaces of the Moon, Mercury and Mars. This is partly a function of the shielding effect of the dense Venusian atmosphere and partly due to resurfacing of large regions of the planet during the last 300 – 500 million years. On radar images typical craters have bright rims and extensive bright ejecta. In some instances the ejecta debris appears to have entrained the atmospheric gases and flowed across the surface for considerable distances. Strange dark haloes also occur; these are believed to represent what appears to be uniquely Venusian atmospheric cratering, whereby an object too small to reach the surface compresses the atmospheric gases against a region of the surface immediately beneath itself, sweeping the ground clear of debris.

IMPACT CRATERS ON VENUS

9

Impact craters range in size from over 250 km, down to small crater groupings that are a mere 7 or 8 km across. Smaller impacts produced multiple crater groups, while the presence of dark surface haloes is a manifestation of "atmospheric cratering". The ejecta surrounding Venusian craters often show distinctive fluidal structure that may involve impact melt flows or impact debris flows.

Pioneer-Venus revealed large circular structures on Venus at roughly the same time as images showing crateriform structures were received at Arecibo. The arrival of Veneras 15 and 16, with their greater resolution, aided by stereoscopic coverage, proved once-and-for-all that such structures exist and, although only 40% of the globe was imaged at a sufficiently high resolution, 146 circular structures larger than 8 km diameter were catalogued. By July 1992, Magellan had imaged 89% of the Venusian

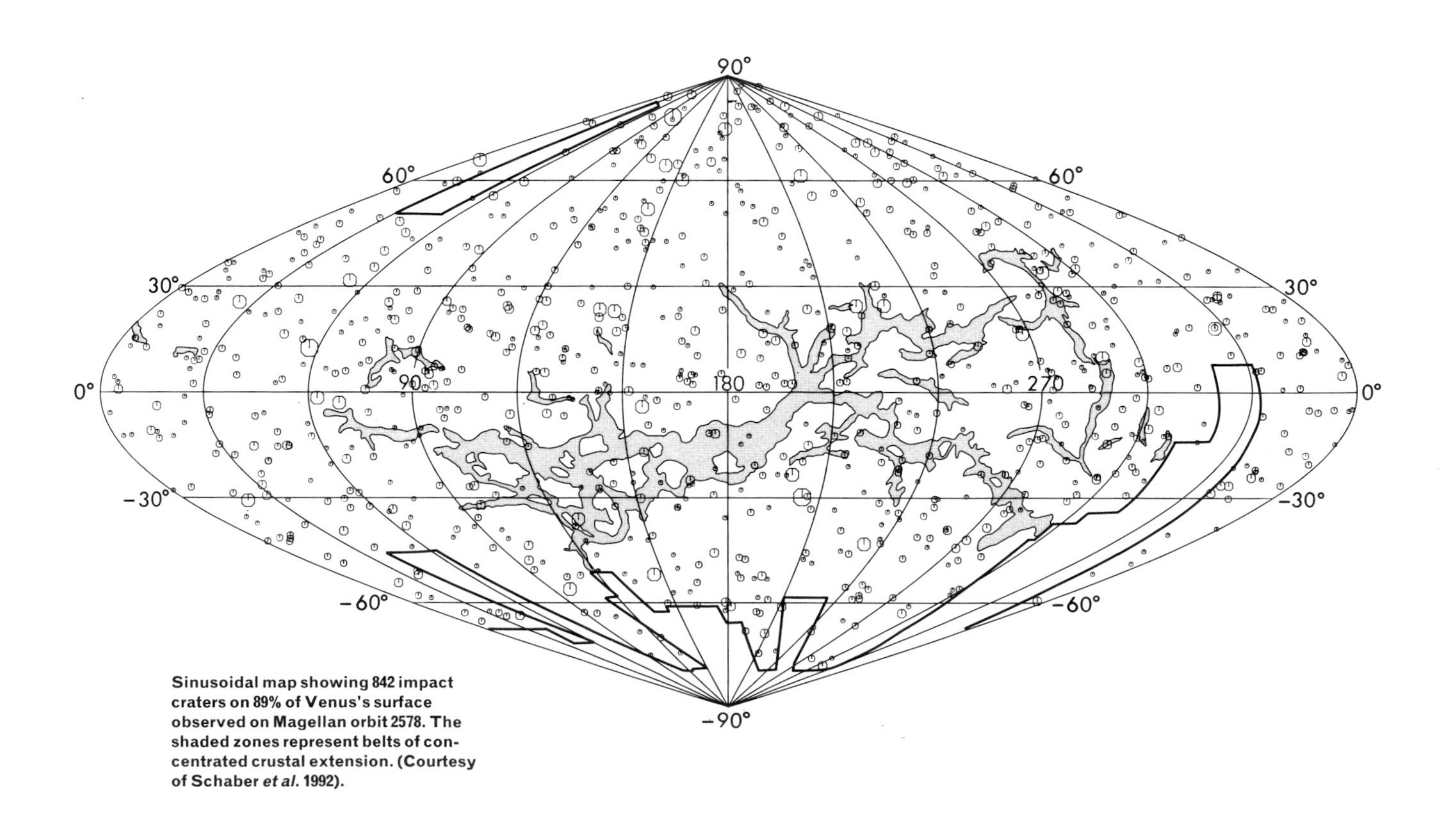

Sinusoidal map showing 842 impact craters on 89% of Venus's surface observed on Magellan orbit 2578. The shaded zones represent belts of concentrated crustal extension. (Courtesy of Schaber *et al.* 1992).

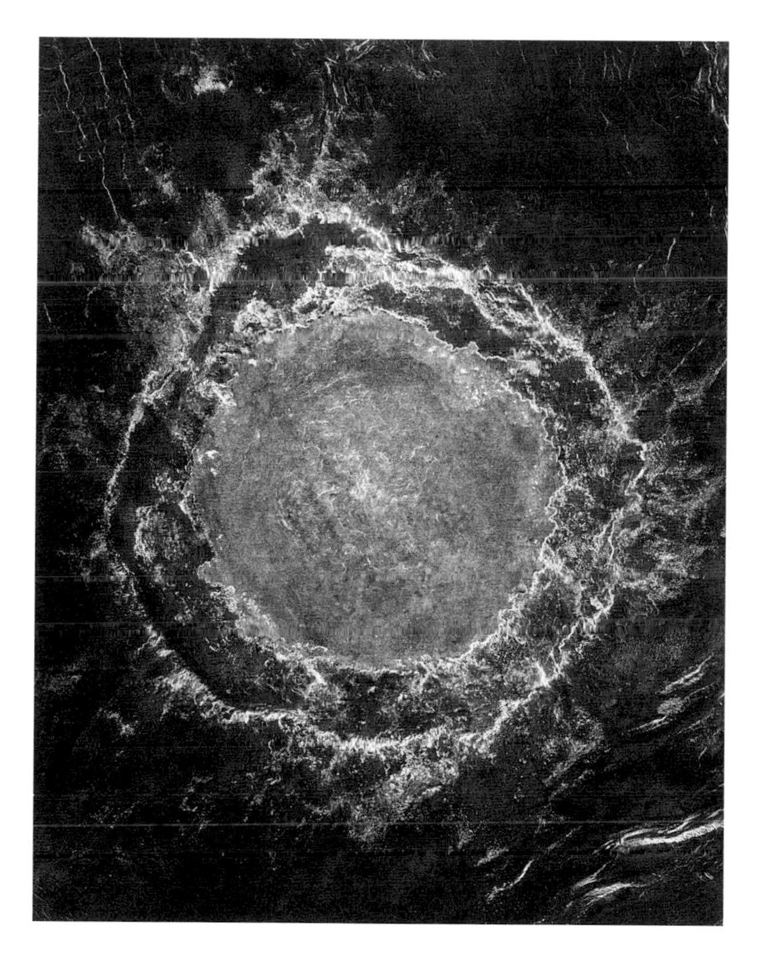

The largest impact crater on Venus, Mead, has a diameter of 280 km and is a multi-ringed structure. The innermost, concentric scarp is interpreted to be the rim of the original crater. The surrounding floor terrace is thought to represent a giant, rotated block that lies outside the original crater cavity.

surface and extended the tally of craters to 842; their diameters range from 1.5 km to 280 km.

The spatial distribution of the craters is remarkably uniform, with over 60% being in a pristine condition; a further 4% are embayed by volcanic flows, while the rest have been affected by tectonism. The apparent concentration of the more modified craters within major fracture-belts and within the highland massifs of the rifted Aphrodite–Atla–Beta–Themis–Phoebe zone, provides evidence that there has been recent volcanic and tectonic activity at a low level. In complete contrast to what is found on the Moon, the spectrum of smaller craters – the numbers of which increase towards the smaller sizes on the Moon – is the reverse of this. This is a direct result of filtering of the smaller (potential) impactors by the dense Venusian atmosphere.

Crater morphology

Most Venusian impact craters are virtually pristine and have sharply-defined rims. Craters are encircled by well-preserved ejecta deposits. There appears to be a progression of morphological characteristics which is size-related: thus the largest structures are multi-ringed, followed at smaller diameters successively by structures with double rings, craters with one ring, craters with a central peak, craters having structureless floors and, finally, irregular or multiple craters. The final group represents a divergence from the familiar lunar pattern, where the smaller craters are simple, bowl-shaped, and circular.

The materials of crater rims characteristically have a strong radar-backscatter and therefore a bright signature; near-rim ejecta have a hummocky appearance and fan out into lobes or flaps. Such a pattern occurs around the 110-km-diameter crater Stanton, one of the larger craters on the planet. Further out, while radar-bright patches still occur, ejecta tend to exhibit lower backscatter and darker signatures. Generally speaking rim ejecta facies extend outwards for about three crater radii. The largest crater, Mead, has a diameter of 280 km and is a beautiful multi-ringed structure. Inside the inner ring the radar signature is of intermediate brightness, indicative of fairly rough materials; these become more hummocky towards the central region. Similar materials characterize the region outside the second ring, but beyond that, and between it and the third ring, the materials have a low backscatter, implying that smoother units dominate.

The effect of Venus's atmosphere

There are significant differences between Venus and those other planets and moons which have extremely tenuous or non-existent atmospheres. It can be shown that the most probable entry velocity of meteoroids into the Venusian atmosphere is between 18–25 km s^{-1}, and recent research indicates that, under such conditions, at a height of around 80 km from the ground, a rapidly-moving meteoroid will deform inelastically, while as it approaches to within 30 km, deformation approaches hydrodynamic conditions. In consequence, projectiles become

Modifications to craters. In contrast to the Moon, only a very small percentage of craters appear to be undergoing removal, volcanic inundation or tectonism. However, the total number of craters on the planet's surface is small compared with the lunar sample and considerable caution must be exercised in interpreting the data. The limited crater sample predictably is a function of atmospheric filtering but account must also be taken of the distinct possibility that resurfacing events may have modified the population, before one can arrive at a credible interpretation of what the data mean in terms of surface ages.

When the locations, relative sizes and types of crater are plotted globally, first impressions are that the distribution is entirely random . Visual inspection reveals that although pristine craters with radar-bright ejecta blankets typify large areas of the equatorial plains, approximately 17% of the population can be considered to have undergone modification. Modified craters have either been partly filled by volcanic plains or had their ejecta deposits (i) embayed by lava flows (ii) or tectonized – that is, affected by fractures and faults which transect some or all of their walls, floors and ejecta blankets. The distribution of such modified craters is not spatially random. Specifically, the volcano–tectonic zones of Beta, Atla and Themis Regiones are not only deficient in pristine craters but actually have an excess of modified and tectonized craters. This inevitably implies that there is a non-random distribution of such craters on the surface of Venus, and that a degree of resurfacing has occurred. Statistical analysis of crater size, frequency and distribution confirms that the western part of Aphrodite Terra – a region where volcanism is relatively lacking – shows evidence for craters having been removed by intense tectonism, while over the western portion of the Beta–Atla–Thetis Regiones zone, both tectonism and volcanism have worked together to eradicate a proportion of the record. An important question is when this may have occurred, and whether or not it was a catastrophic global event.

flattened rather than rounded, in this way differing from similar objects which impacted the Moon. The smallest craters which can be created, at least for low-strength meteoroids, are in the size range 7–13 km at zero altitude, and between 6 and 11 km at around 10 km. The smallest craters whose numbers are not affected by the filtering process would be approximately 32 km across.

Using this as a basis, it is then possible to compare the size-frequency curve of craters ≥ 32 km diameter for Venus with that of the Moon, to give an idea of the resurfacing history of the planet. Very evident is the similar size-distribution on both worlds. However, on Venus there is a much lower crater density – a very significant difference indeed. On the basis of all of the tests so far conducted, an average surface age of around 500 million years has been derived, implying that the latest resurfacing event must have occurred as recently as this. In some areas an even lower number of craters suggests locally younger ages, as low as 300 million years. However, it is possible that this is an underestimate of the real age since cratering efficiency on Venus may be less than on airless worlds.

Crater ejecta

Because of the density of the atmosphere, a Venus approaching projectile will partition a proportion of its kinetic energy directly to the atmosphere prior to hitting the ground. In the much-quoted Tunguska event, where an incoming comet catastrophically disrupted above the Earth's surface, most, if not all of the energy apparently was partitioned to

Sinusoidal map showing those impact craters which have been modified by lava embayment (small squares) and by fracturing (big squares). (Courtesy of Schaber *et al.* 1992)

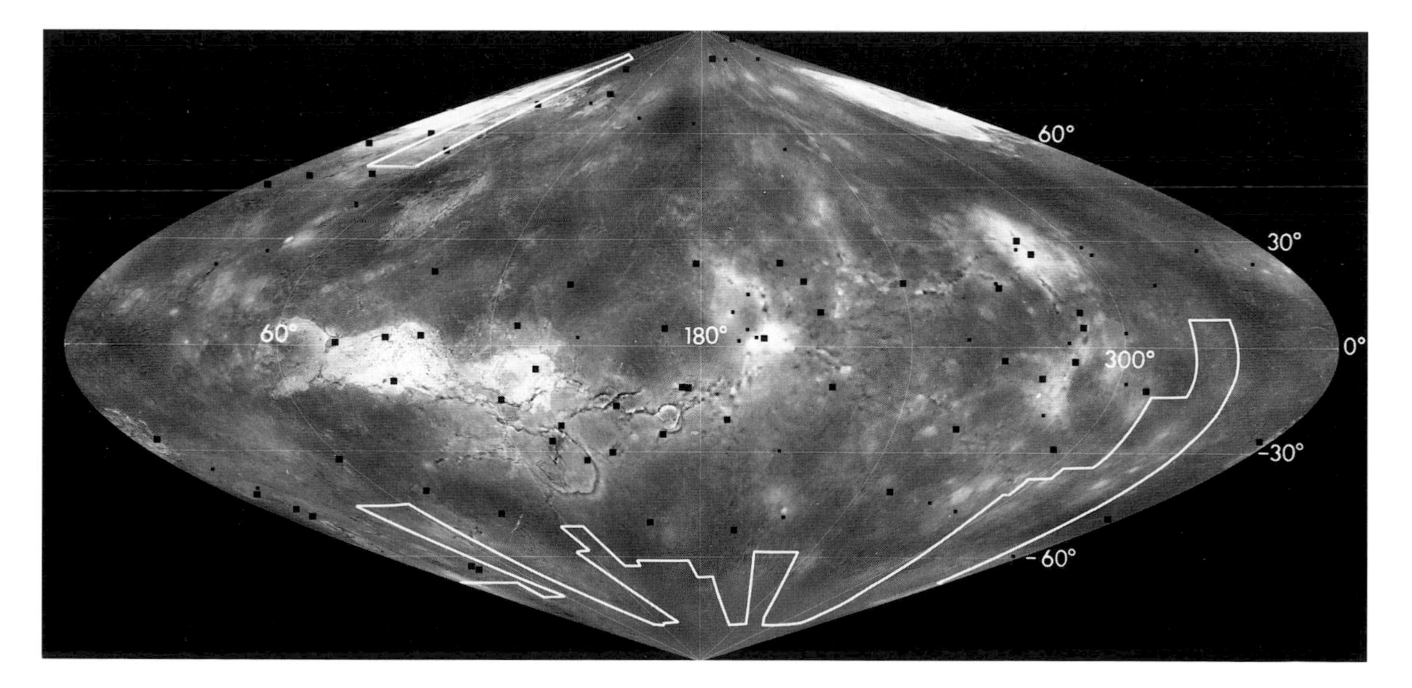

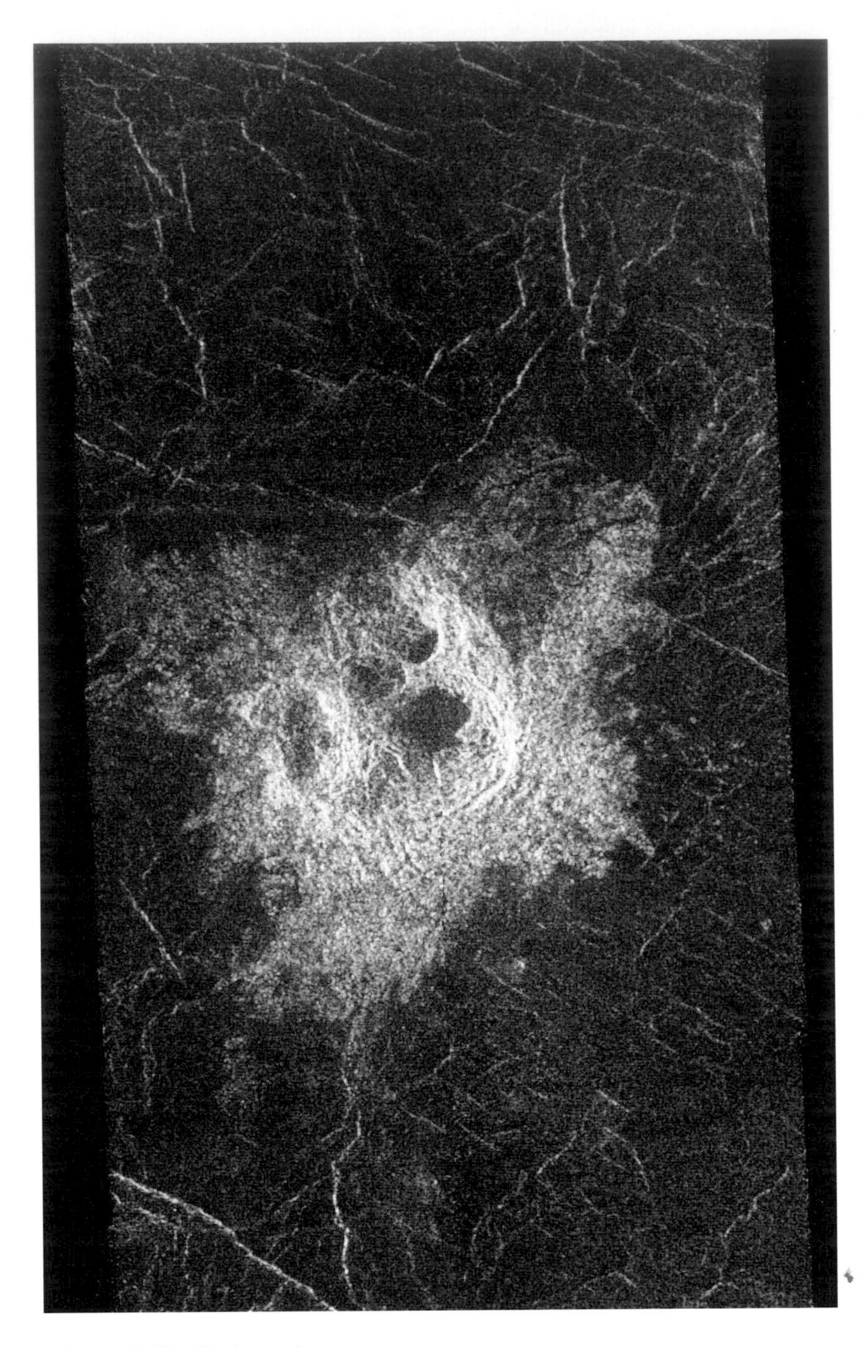

A 40-km-wide Magellan image showing an irregular (composite) crater 14 km in diameter. Smaller bolides break up in the dense Venusian air, producing multiple cavities with a complex outline.

the atmosphere, the shock wave flattening trees over a very large region.

The energy released by a non-disrupted meteoroid upon impact is partitioned to the atmosphere by rapidly-moving ejectamenta, a fast-expanding cloud of vapour, frictional drag, and from highly turbulent disturbances within the atmosphere induced by the outward-moving curtain of ejectamenta. This has far-reaching consequences, both for the emplacement of ejecta and for its effect upon the planet's surface. Because a significant amount of energy is partitioned to the atmosphere during the passage of a projectile, the gas flow associated with its passage through the dense atmosphere will imitate a strong explosion. Furthermore, a low residual gas "density-tube" will form along the trajectory, its radius being large compared with the diameter of the resultant crater. The effects of this are significant, and ejecta entrained in the 2–3-radius-wide region around the focus of a crater will not have been distributed by "normal" atmosphere, which would generate lunar-like patterns. Thus the boundary of the flow of dense gas probably corresponds to the inner zone of hummocky ejecta.

Upon impact, a huge amount of rising turbulent hot gas not only disturbs ejecta deposition but also transports large volumes of dust into the upper atmosphere. Because of the great density of the Venusian air, a very considerable pressure differential will be set up on either side of the advancing curtain of ejectamenta, giving rise to extreme turbulence. This has the ability to scour the region within 4–6 radii of the crater rim, creating a coarse lag deposit that is characterized by a radar-bright signature.

Perspective view of three large impact craters in Lavinia Planitia.

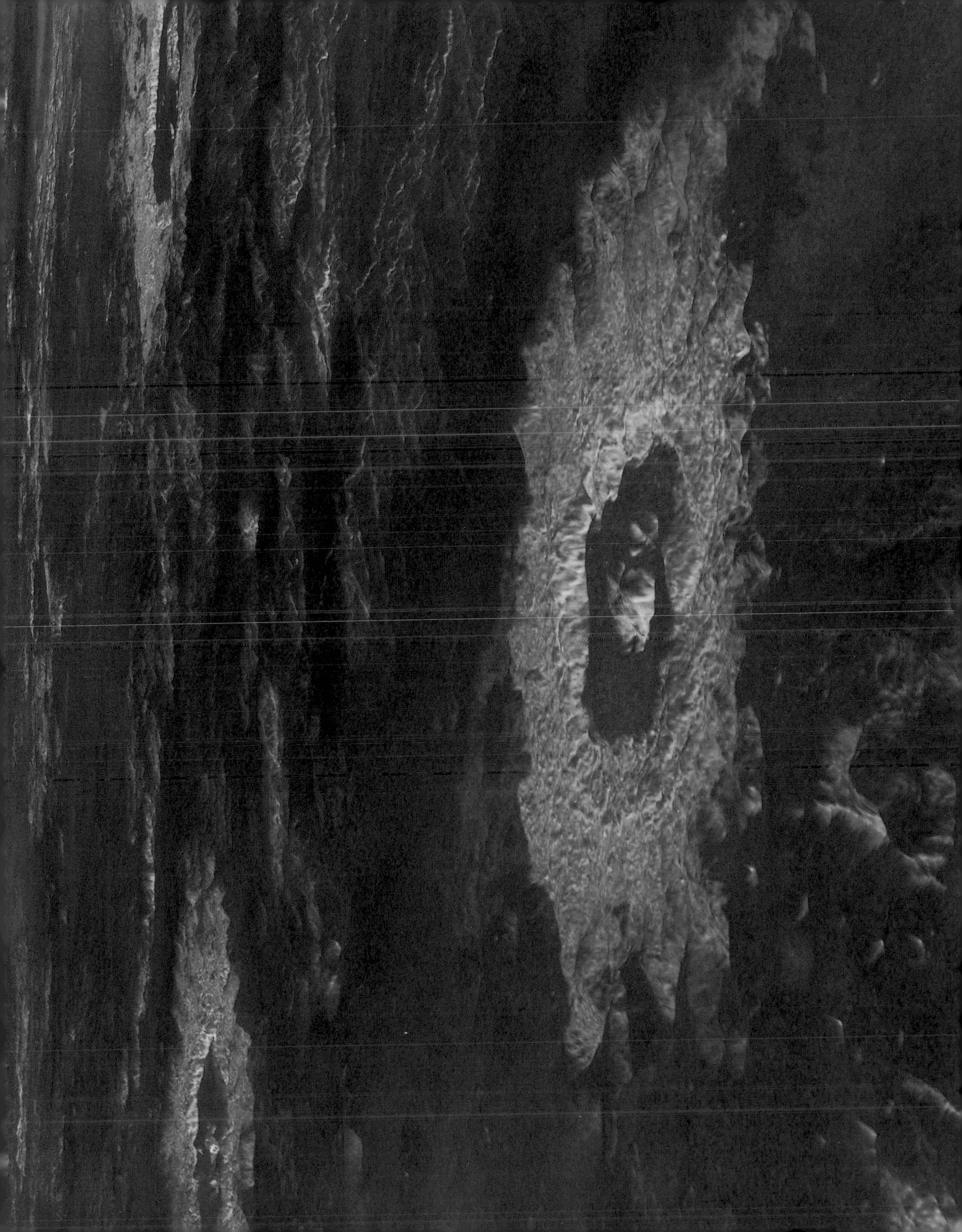

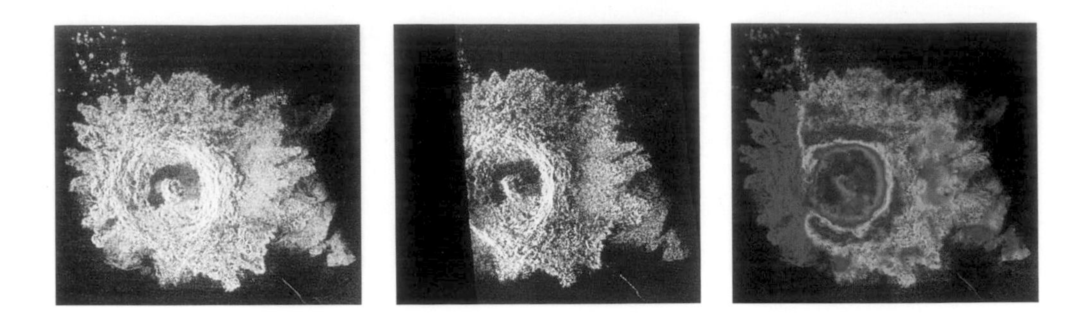

This set of radar images demonstrates how stereo imagery can be used to produce high-resolution topographic information. The two images on the left, of the 24-km-diameter crater, Riley, are from Magellan's first and third mapping cycles, respectively, while the third, rectified for distortion, is a colour-coded altimetric map generated from the stereo pair to its left.

The Venusian crater, Isabella, 175 km in diameter, showing the two extensive flow-like features which extend away from the crater. The end of the more southerly flow partially surrounds a pre-existing 40-km-diameter volcanic shield, while the southeastern flow shows a complex pattern of flow lobes and channels.

Immediately behind the advancing curtain of ballistically-ejected material, separation of coarse and fine debris is likely to occur, with the result that runout of fine-grained particles entrained in the outwardly flowing gas goes well beyond the inner hummocky zone. It is also probable that the shock-wave-induced flow of finer ejecta beyond the hummocky zone may resemble that in turbidity currents, in which case such flows would have the power to erode sand-sized fragments, pebbles and boulders within this zone, giving rise to the distinctive radar-dark collars seen.

Another phenomenon arises due to the slowing down of meteoroids as they descend through the atmosphere, and of any vapour and melt generated during descent. Thus, highly mobile deposits may be emplaced well before impact and crater excavation, these being due to the collapse of the column of low density gas that follows in the wake of the bolide. The surface morphology of such impact-generated flows, as revealed by Magellan imagery, suggests that while many are comparable with terrestrial turbidity currents, others were distributed by laminar flow.

THE HIGHLAND MASSIFS

10

Although of small areal extent, the planet's highland massifs are some of the most complex geological structures in the Solar System. They reveal a long history of deformation which has been interspersed with volcanic activity and cratering. The unique tessera massifs have no terrestrial counterparts.

The Earth's continents account for approximately 30% of the planet's surface area. In contrast, the Venusian highlands – the most continent-like regions of Earth's sister planet – cover a mere 10–15%. The Venusian highlands rise on average 4 to 5 km high above mean planetary radius (MPR) and, although of small areal extent, are particularly intriguing, not least because of their geological complexity. And it is particularly unfortunate that, up until now, no chemical analysis of highland rocks exists. However, it is generally assumed that, like terrestrial continental rocks, their composition is more "evolved" than those of the basaltic lowlands analysed by Venera and Vega probes; this means that they probably are richer in silica and less dense than the plains materials. This would help them to stand "proud" of the surrounding plains units.

Types of highland terrain

There appear to be three kinds of highland massif. First there are *Beta-type highlands* – broad topographic rises showing signs of moderate extensional deformation, shield-style volcanism akin to that of Hawaii, and deep levels of isostatic compensation. These, in essence, are large volcanic rises. Then there are *Ovda-type highlands* – topographic plateaux distinguished by intense deformation, but which show more limited signs of volcanism and shallower depths of isostatic compensation (<100 km), and, finally, *Lakshmi-type highlands* (Ishtar Terra is the only one) – an elevated plateau with encircling narrow mountain belts, a moderate development of volcanism and deep compensation depths. The three types appear to have different origins and may be of diverse age too.

The equatorial highlands

The most extensive highland region on Venus is Aphrodite Terra, which stretches along the equator between the 45° and 210° meridians. Aphrodite exhibits considerable variation in physiography, which is paralleled by increasing apparent depths of compensation from west to east. Western Aphrodite is built from the massifs of Ovda and Thetis Regiones which rise between 3 and 4 km above datum, and extend for a distance of about 6000 km. Magellan has shown that these western

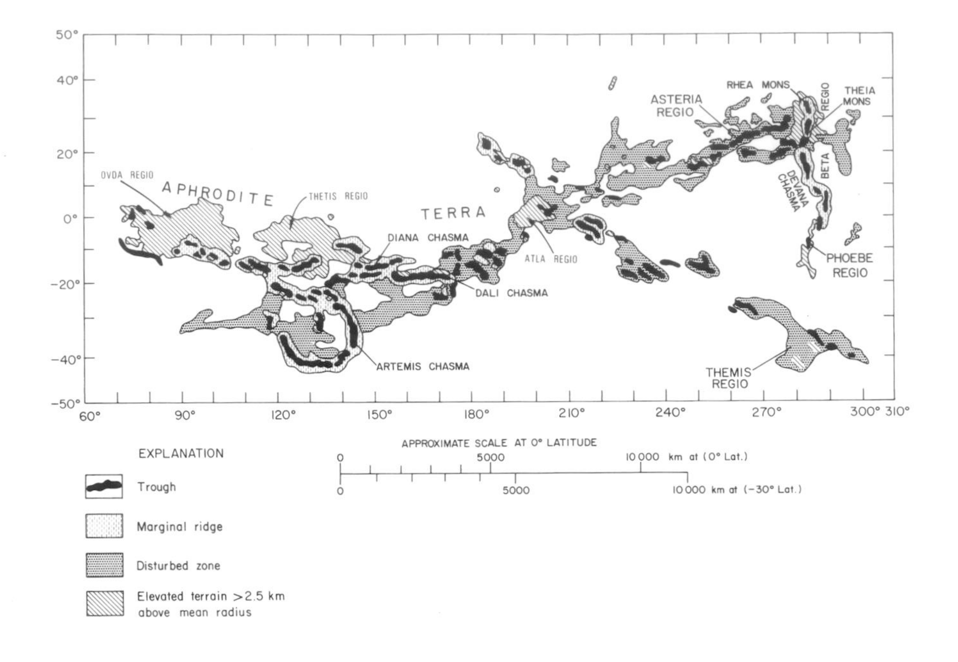

Annotated map of the components of Aphrodite Terra (After Schaber 1982).

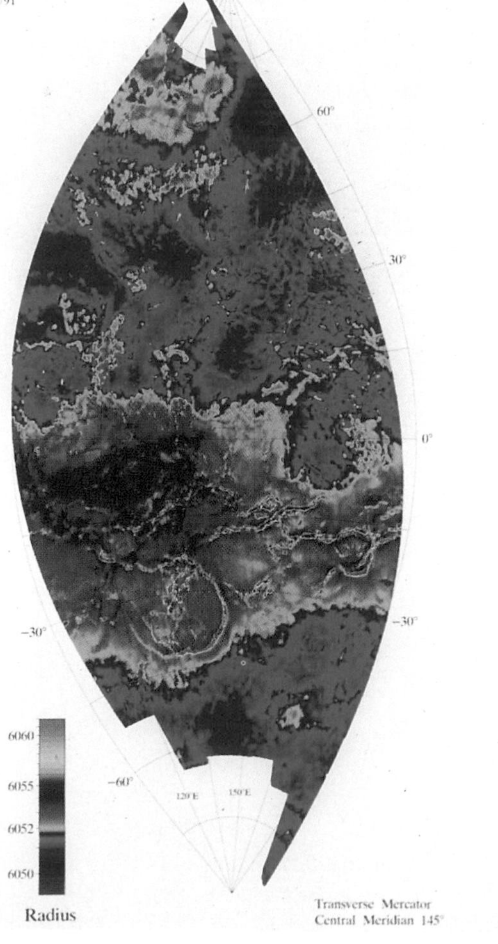

A false-colour representation of the surface relief across a part of Aphrodite Terra. The distance from top to bottom is 19,000 km. The large curvilinear feature is Artemis Chasma.

massifs are cut by rift faults, most of which trend WNW–ESE. Also, both massifs are deformed in a very complex way and their margins highly fractured. The interiors of both highland blocks are a complex of blocks, broad domes, linear ridges and troughs with predominantly radar-bright signatures indicative of rough surfaces. However, the floors of many grabens (small rift faults) are covered by radar-dark materials presumed to be volcanic flows. This type of structure is typical of Venusian terrains termed *tesserae*, also known by the less pleasing term *complex ridged terrain* (CRT for short).

Centred on the highest point of Ovda Regio is a gravity anomaly which suggests an apparent compensation depth of 70 ± 7 km. The gravity signature of Thetis is considerably more complicated, and is not entirely understood. Both regions of tessera structure must be being supported from movements within the underlying mantle. Similar terrain extends to the west of Ovda for at least 1500 km and may well stretch for a further 500 km, making it even more extensive than Fortuna Tessera, the tessera region extending eastwards from Maxwell Montes. Some planetary scientists are of the opinion that the tectonic framework exposed here

Magellan radar image of Ovda Regio, a compilation from all three 8-month-long mapping cycles. It covers an area of 6300 km square. Ovda rises 3 km above the mean surface level and is made up from complex deformed terrain called tesserae. The effects of deformation are particularly well seen along its margins, where curvilinear ridges and troughs have developed. The huge impact crater, Mead, lies to the northwest of Ovda's western extremity.

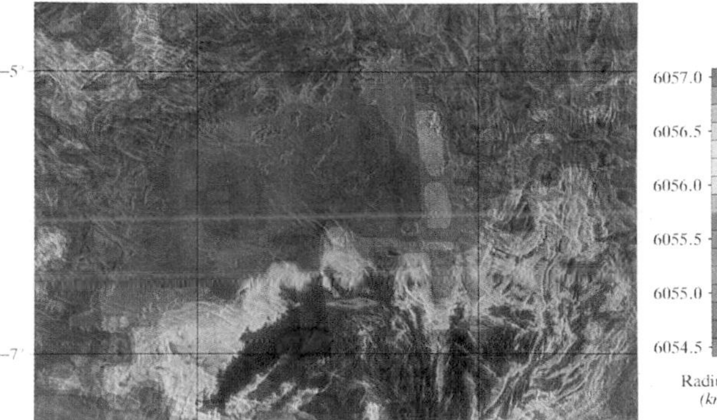

Topographic map of the eastern part of Ovda Regio. A thick flow of lava extends about 160 km east–west near the centre of the image. The topography indicates that its central region lies nearly 1 km below its edges.

is consistent with an early phase of crustal shortening (compression) which was followed by more modest extension; others favour the idea that the Venusian mantle undergoes downward circulation beneath these plateau-like highlands.

To the east of Thetis the landscape seldom rises above 1.5 km and this is so for a further 3000 km. Beyond this point Aphrodite's central region is characterized by a series of ENE–WSW trending troughs and ridges, particularly imposing being Diana and Dali Chasmata. Within these rifted zones elevation differences of 7 km are not uncommon, and this over horizontal distances of only 30 km, meaning average slopes must be of the order of 13°.

The eastern part of Aphrodite is occupied by Atla Regio with an average elevation of 3 km, which is a broad dome on whose surface are numerous rifts and volcanic centres. Above its surface rise the imposing volcanic peaks of Ozza and Maat Montes. Ozza Mons is 6 km high and 300 km across, while Maat Mons is 9 km high and 200 km across and has several summit depressions, or calderae. A third similar but smaller volcanic centre, Sapas Mons, is situated further to the west. Radar images are characterized by radiating, lobate, volcanic flows giving rise to radar-bright signatures indicative of rough

A computer-generated perspective view of the highland tessera of Ovda Regio.

lava surfaces. Clearly, because Atla is dominated by volcanic deposits it differs significantly from the highland regions of Western Aphrodite. It is not a tessera massif but a large volcanic rise.

Imposing though the volcanoes undoubtedly are, equally prominent are the huge rift faults. The 1000-km-long Ganis Chasma, which connects Maat Mons to Nokomis Montes, is composed of numerous individual graben 1–10 km wide, has a maximum width of 300 km and shows relief differences of between 1.0 and 1.5 km with respect to the adjacent plains. This is just one of several such structures. Individual faults intersect one another, forming a branching network. South of central Aphrodite Terra lies the curvilinear Artemis Chasma which appears to be a massive corona structure rather than a part of a regional fracture system.

Tellus and Bell Regiones

Three thousand kilometres to the north of Ovda Regio is the plateau of Tellus Regio which rises to around 2 km. With fractured margins and a radar-rough surface, it is similar to Ovda and Thetis, even in having embayments of smoother radar-dark plains within its complex ridged terrain. However, although there is a weak positive gravity anomaly associated with its eastern part, it

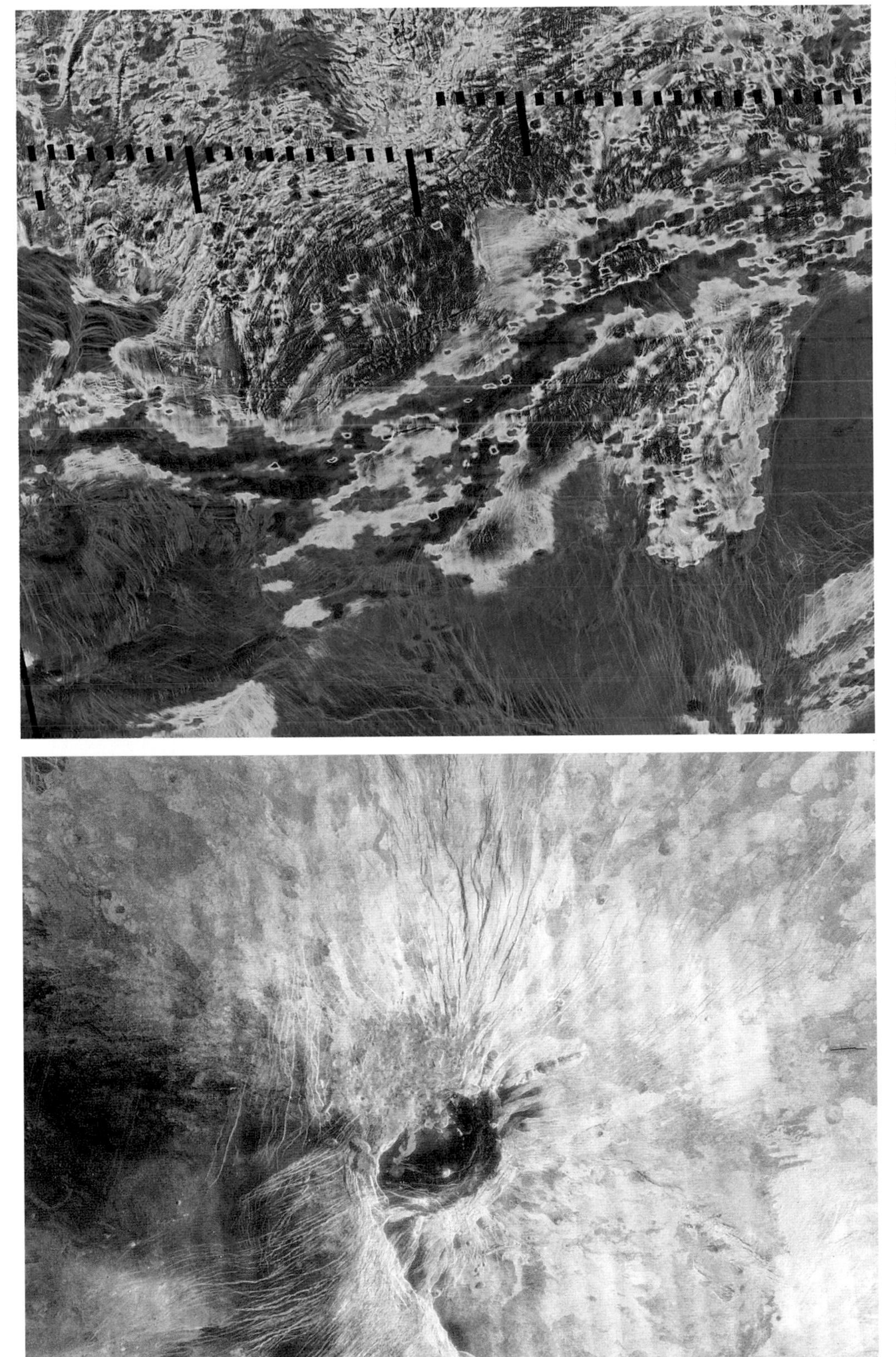

A false-colour image of a part of Aphrodite Terra, showing the prominent rift fault that separates the highland massif from the adjacent plains. The rift shows up as the dark-blue strip in the lower part of the image.

The summit region of Ozza Mons volcano, Atla Regio. The radar-dark summit plateau and complex fracture system are clearly visible. This highland is a huge volcanic rise.

differs significantly from other such highlands in that the major anomalies in its vicinity are strongly negative. Bell Regio, which sits west of Tellus, rises to between 1.5 and 2.0 km, and hosts a 500-km-diameter volcanic shield called Tepev Mons and the large corona, Nefertiti. Like Atla, this is more by nature a volcanic rise than a plateau structure and both Bell and Tellus Regiones may sit above active *mantle plumes*.

Beta Regio

This large highland measures 2000 km × 3000 km and is split by the prominent northerly-trending rift fault, Devana Chasma, which is characterized by at least 6 km of relief in places, and is a major regional fault complex comparable with the Earth's East African Rift. Beneath it is a large free-air positive gravity anomaly, sug-

Composite image of the northern part of Beta Regio, showing the rifted tessera massif of Rhea Mons. Radar-dark plains lie on either side, while midway along the rift can be seen the deformed impact crater, Somerville.

gesting an apparent depth of compensation of around 300 km.

The mountain of Theia Mons lies at the southern end of Beta, and is a huge volcano that has been superimposed on the rift. Theia has a diameter of 350 km and an altitude of 5 km, and is the focus of numerous long lava flows which partly fill in the downfaulted block. Theia also sits at the junction of features with different tectonic trends, the most obvious being the westerly-trending Hecate Chasma. Not surprisingly, this has led some scientists to draw parallels with terrestrial rifts, where faults are propagated and may eventually link up between individual regions of hotspot-related magmatism.

Rhea Mons, located in the north of Beta, formerly was interpreted from Arecibo and Venera images to be a volcano, like Theia Mons. However, Magellan revealed the

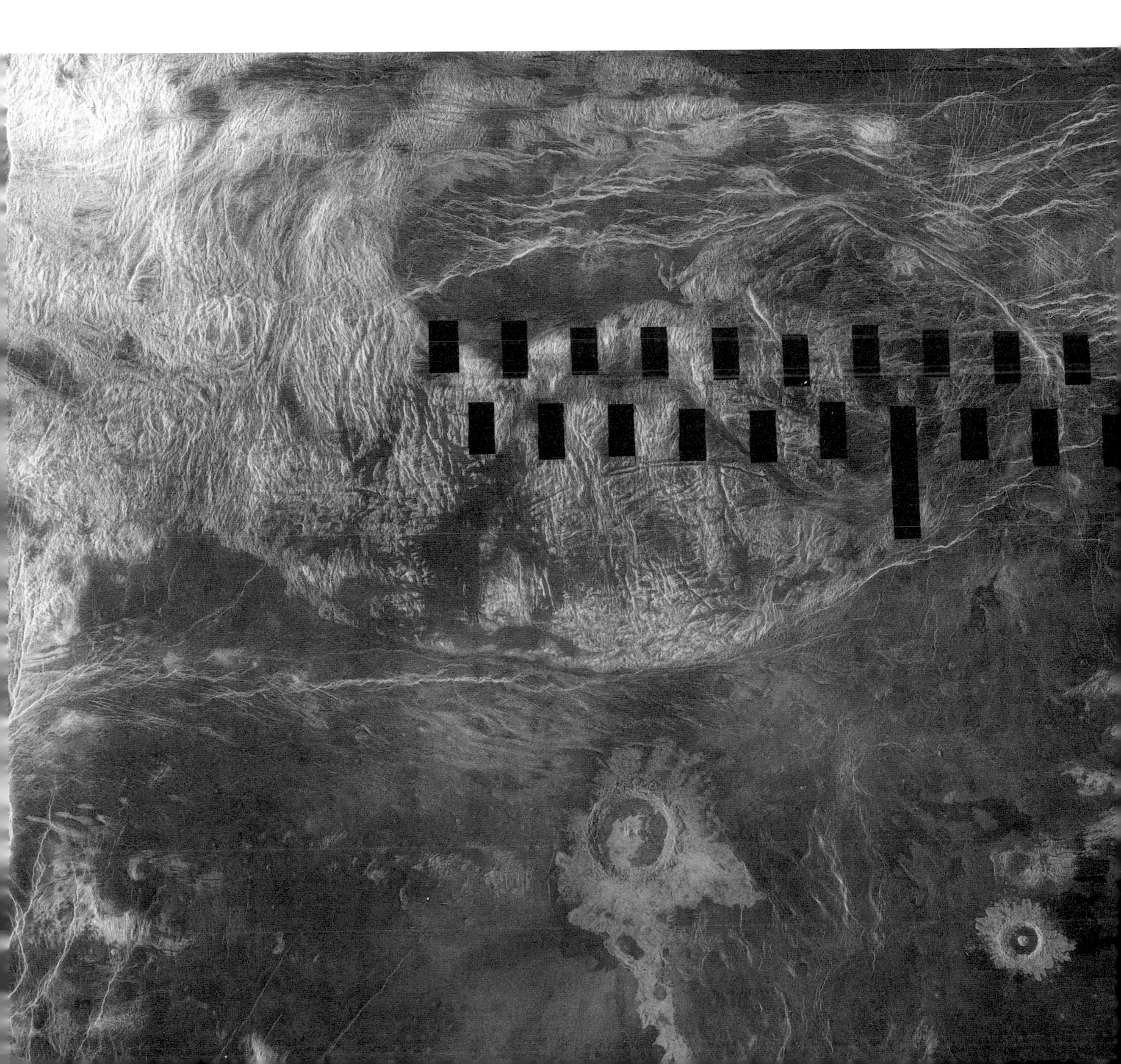

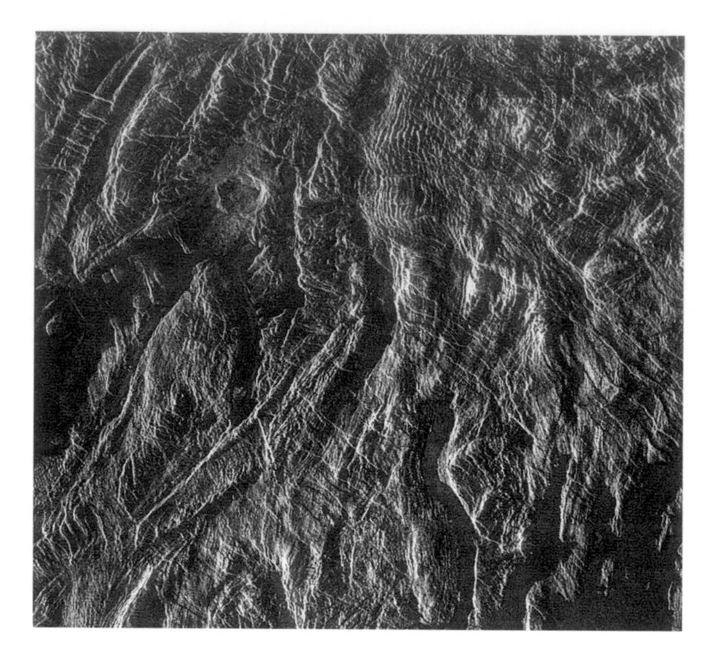

The northern boundary of Alpha Regio, showing criss-crossing graben faults which pass onto the radar-dark plains units.

The region of Beta Regio, showing the huge volcanic construct of Theia Mons and the rift system of Devana Chasma. The northern massif, Rhea Mons, appears to be a rifted-apart tessera massif.

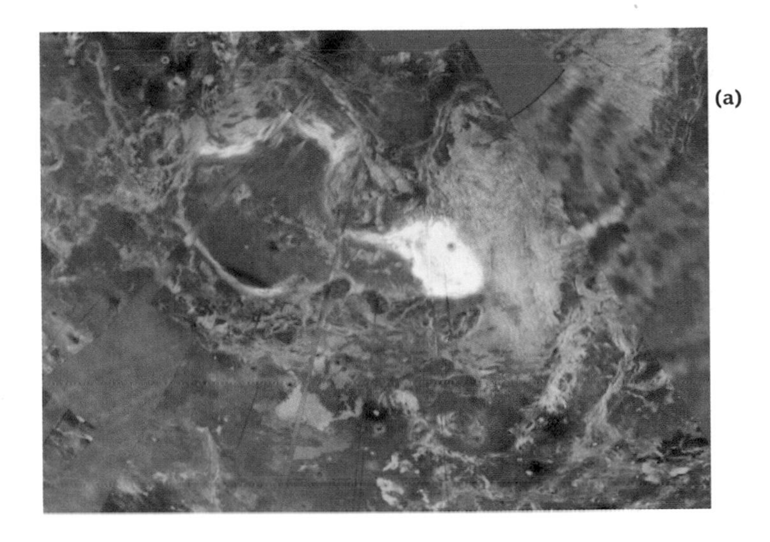

(a)

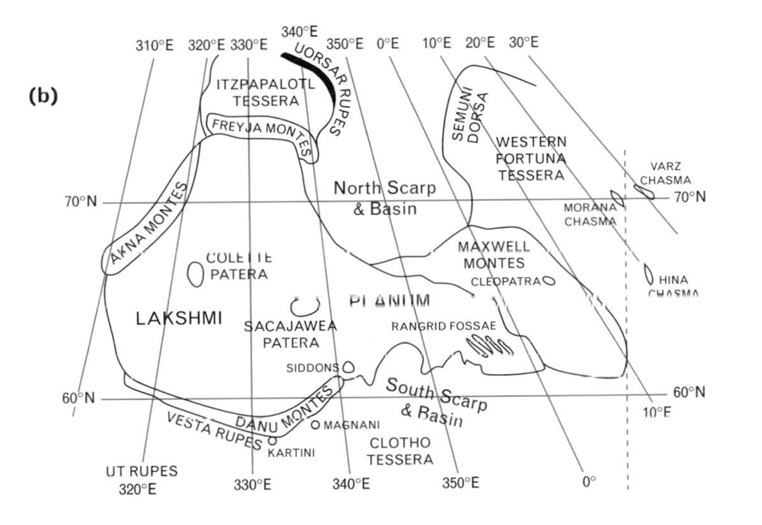

(a) The northern hemisphere of Venus showing the highland massif of Ishtar Terra. The pale-coloured massif is Maxwell Montes. To its west is the huge plateau of Lakshmi Planum, and to its right the extensive region of Fortuna Tessera. (b) Annotated sketch map of Ishtar Terra.

most elevated part to be an uplifted tessera block which, at its summit, is mantled by smoother volcanic deposits. Near Rhea the fault trough is relatively deep (>2 km) and narrow (80 km), whereas in northern Beta the trough is broader (130 km wide) and there are higher-standing blocks in the rift centre. To the north of Beta is a chain of large coronae linked by faults; these are younger than fractures splaying out northwards from Devana Chasma, and may represent a later phase of activity in this region. To the south of Beta Regio the structural line is continued first through Phoebe, then Themis Regio and then Tefnut Mons which also may have volcanic origins.

Ishtar Terra

The highest point on Venus (12 km), situated within Maxwell Montes, is situated towards the eastern end of this Australia-sized highland. It forms the core to a massive elevated region with a very steep western flank. The most recent Magellan data indicate that the southwestern flank of Maxwell has a slope of 35°! Eastwards, however, it drops more gently towards the highly deformed area known as Fortuna Tessera which extends for a distance of nearly 2000 km.

The western side of Ishtar comprises the 4-km-high elevated plateau of Lakshmi

Planum, which is bounded by curved belts of high mountains. This covers an area of two million square kilometres – approximately twice that of the Earth's Tibetan Plateau. The surface, somewhat featureless at medium resolution, is broken by two major shield volcanoes with deep calderas. These two landforms, called Colette and Sacajawea, dominate central Lakshmi Planum. The encircling mountain belts of Danu, Akna, Freyja and Maxwell Montes form the north, west, south and east boundaries respectively, while the steep escarpments of Vesta Rupes, to the southwest, and unnamed scarps to the southeast and northeast, sharply define its limits. External to the encircling mountain belts are elevated regions of tesserae which extend downslope over distances of between 100 and 1000 km.

Evolution of Ishtar Terra

The great height of central Ishtar has led many scientists to suppose that it is a region of vigorous (possibly contemporary) crustal convergence, that is, a region of Venus where areas of the lithosphere are being driven together. Exactly why this is so is not entirely understood; some believe we have mantle downwelling below Ishtar,

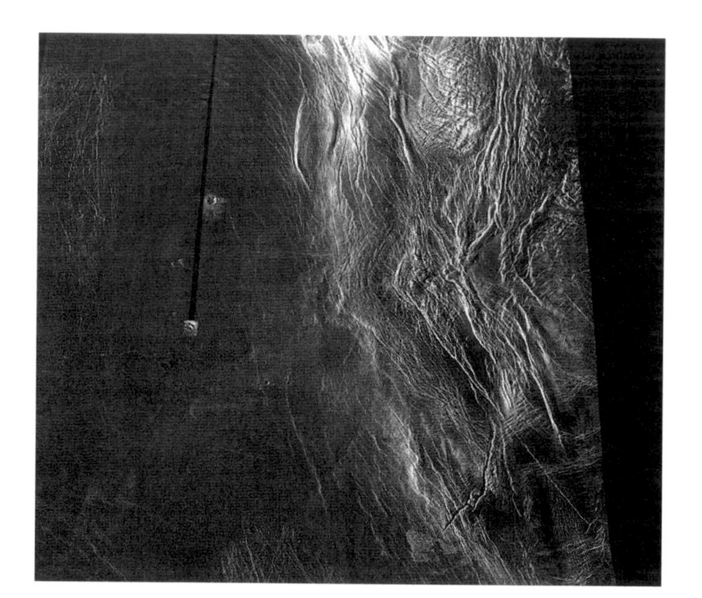

The southern part of the mountain belt of Freyja Montes is seen here on the righthand side of the image. North–south ridges are believed to be due to crustal folding or due to low-angle thrust faults. To the west lie the volcanic plains of Lakshmi Planum.

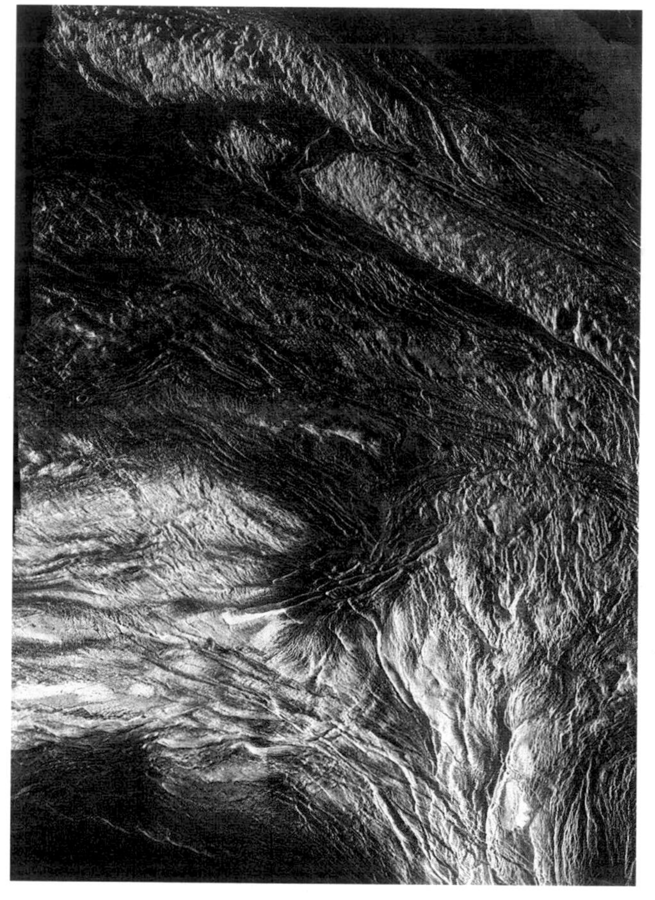

Details of the complex deformed belt of Freyja Montes, with Lakshmi Planum to the south. Note the sinuous lava channels that have flowed across the intervening radar-dark regions.

Colour-coded thermal emissivity image of a part of Maxwell Montes and Fortuna Tessera. In the left-centre of the image is the structure Cleopatra, by some believed to be an impact crater, by others a volcanic caldera.

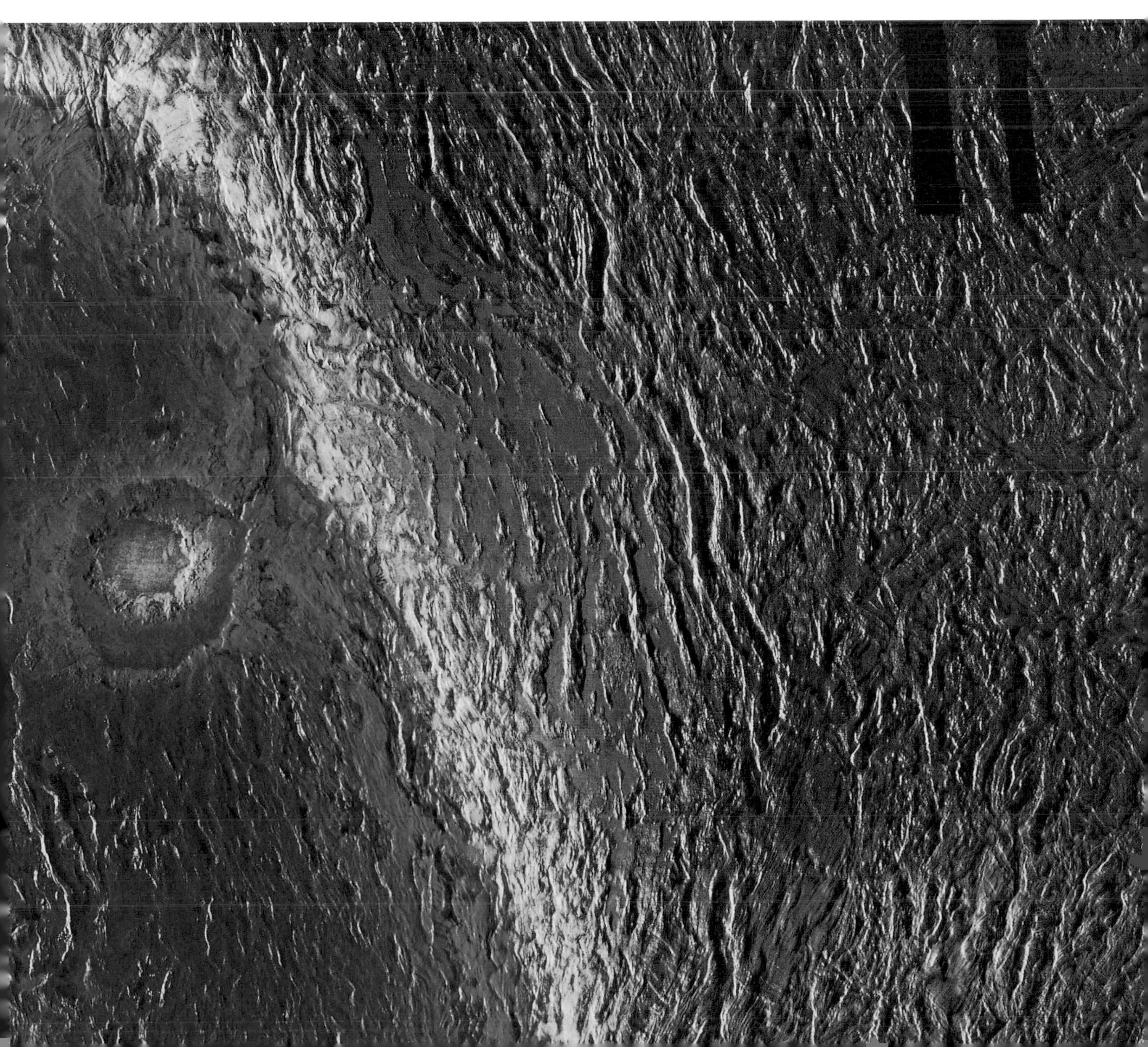

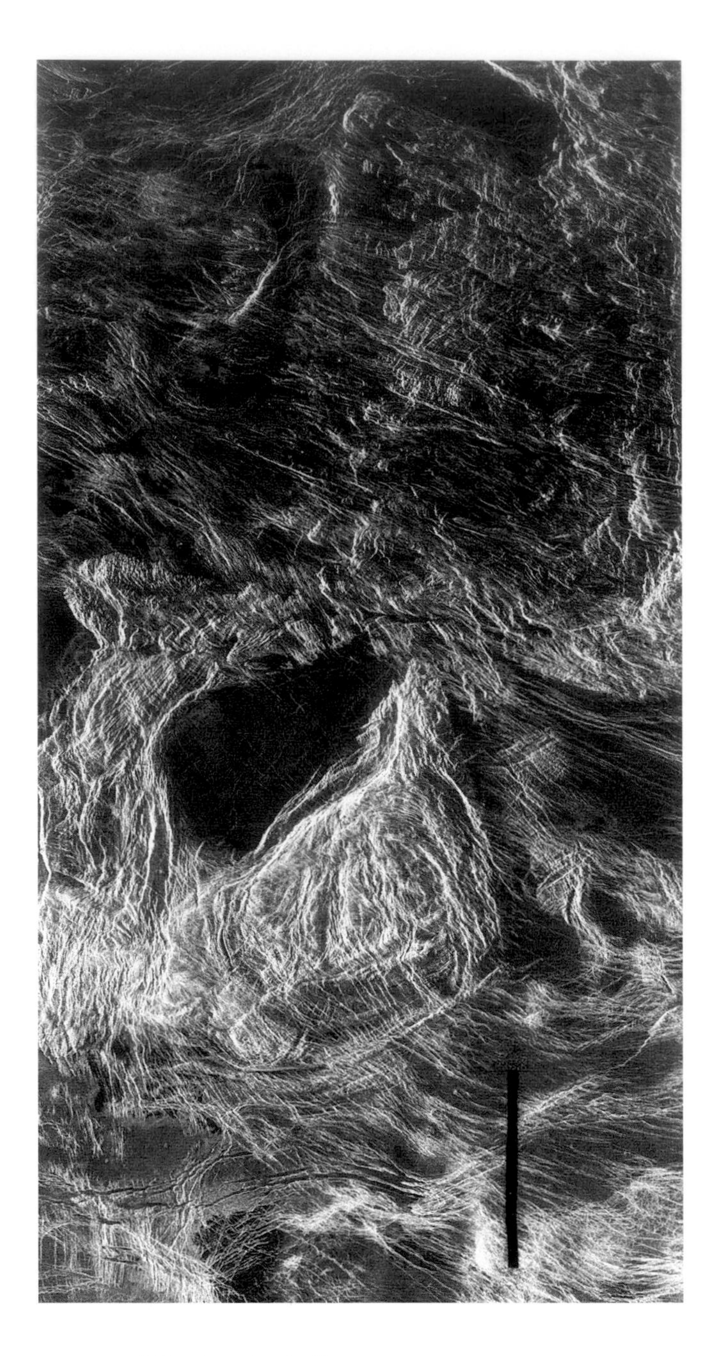

Ishtar Terra, showing the western part of the region of Clotho Tessera. This is dominated by intersecting ridges and valleys, while domes, pits and smooth dark plains indicate that volcanism has played an important part in moulding the intervening regions.

others consider it more likely that there is upwelling in places and downwelling elsewhere. What is likely is that beneath Ishtar we have cool, thickened crust that is somehow able to remain proud of the mean planetary radius. Such a situation can be achieved, in one way, by it being buoyed up by an actively-rising region of upwelling mantle, or it could be rucked up by mantle movements that are driving (or have driven) slices of cool crust together. Another interesting possibility is that the variable trend of the structures within individual tessera components of Ishtar Terra may mean that the various components of this complex highland region have been moved laterally, possibly also being rotated, from their original positions during geological time, by persistent or spasmodic mantle forces. If this were eventually proved to be the case, then this unique highland region would probably be the closest Venusian analogue we have for terrestrial-type plate movements associated with continental collision. Clearly much more research needs to be completed before the true history of Ishtar can be revealed; the above idea is at best, highly speculative.

THE HISTORY OF VENUS

11

The geological history of Venus is, as yet, only partly understood. The impact record suggests that the present surface is, on average, only half a billion years old, implying a global resurfacing event at that time. The highland massifs may represent remnants of older terrains. The distribution of major volcanic features indicates that Venus loses its internal heat, not via the sutures in lithospheric plates, but by plume activity.

The very early history of Venus doubtless followed a similar pathway to that of the Earth. Thus, following accretion from planetesimals and a gradual increase in mass, internal heating gradually provided sufficient heat for the melting of metallic iron and its sinking to form a metallic core. Subsequently, further differentiation generated a mantle and crust. Since the mean density of Venus is virtually identical to that of Earth, its core, mantle and crust probably give a density profile rather similar to the Earth's.

Today the atmosphere of Venus is very different from that of the Earth and undoubtedly very different from the original atmosphere that followed degassing of the primordial planet. The process of degassing is believed to have started when the terrestrial planets had roughly one-third of their present masses. As it proceeded, water went into the Earth's oceans, and maybe the same happened on Venus, as there is good geochemical evidence that their original atmospheres were probably rather similar in make up. However, because of Venus's greater proximity to the Sun, it gradually got hotter. The response to this was that more water vapour entered its atmosphere to balance out the dissociation of hydrogen and oxygen in the upper layers of the atmosphere by ultraviolet radiation from the Sun. This prevented infrared windows forming with the result that re-radiated solar energy could not escape into space. The heat became trapped and a greenhouse effect was established. Gradually, more and more water was lost and more and more carbon dioxide entered the Venusian atmosphere. This contrasted strongly with what was happening on the Earth, where CO_2 entered carbonate rocks forming beneath the terrestrial oceans.

Because of the very high surface temperatures which prevail due to the long-lived greenhouse conditions, Venus's lithosphere has to be significantly more buoyant than Earth's, and the planet is less likely to have developed terrestrial-style plate recycling processes. Indeed, the distinctly unimodal hypsographic curve of Venus reflects a complete absence of typical terrestrial ocean-floor and continental-interior profiles. The facts, therefore, argue strongly against the notion that seafloor-spreading or processes akin to this can be taking place at the present time. Whether or not the original water of Venus hung around long enough for the

The large ovoid structure in this image is the 500-km-diameter corona, Eithinoha, one of the many such features believed to be the result of mantle plume activity on Venus. The vital role of volcanism is further evidenced by the features of the radar-dark plains, including volcanic flows, lava channels and a variety of other volcanic landforms. That these have partially flooded regions of older, deformed tessera is clear from this image. The age of the surface can only be estimated from the incidence of impact craters, several of which may be identified here too.

process to start is a moot point; because of subsequent resurfacing, no evidence of such a phenomenon can be left.

In the absence of a method of removing thermal energy from the interior by plate tectonics, mantle plume and hotspot activity ought to be more vigorous on Venus than on the Earth. Gravity data obtained by Magellan suggest that highland massifs are supported by low-density roots that penetrate to depths which, in some instances, must lie well below the lithosphere and that the topography is being dynamically supported, presumably by thermal buoyancy in the mantle. Such buoyancy would be provided by upwelling mantle plumes, these *hotspots* supporting the overlying lithosphere.

The abundance of corona-type landforms and larger-scale volcanic rises such as Beta and Atla Regiones undoubtedly is a manifestation of this widespread plume activity. Such activity, well known from terrestrial experience, has the capacity to generate lithospheric swells, major rifting and associated volcanism – precisely what is seen on Venus. The somewhat "spotty" appearance of volcano-tectonic features on the planet, i.e. their non-linear distribution, reflects the escape of internal heat from thousands of irregularly-spaced hotspots.

Ishtar Terra is the highland massif on Venus which many workers have considered most akin to a terrestrial continent. Lakshmi Planum, surrounded as it is by arcuate zones of folded and thrusted rocks and bounded by steeply-sloping scarps, is seen by them to bear some resemblance to terrestrial island-arc regions, with trench-like features and fold mountain belts. The generation of such features is practicable on the Earth, since it has both an atmosphere and a hydrosphere and is not subject to the stifling, dehydrating, greenhouse effect which is a key feature of Venus. Nevertheless, if Ishtar is to be seen as a "continent" in the terrestrial sense, formed by plate processes, it becomes necessary to find some mechanism to differentiate the original mantle-derived materials. Thus far, it is not possible to envisage how this can have been achieved under present conditions; however, should there once have been lower temperatures on Venus, water could

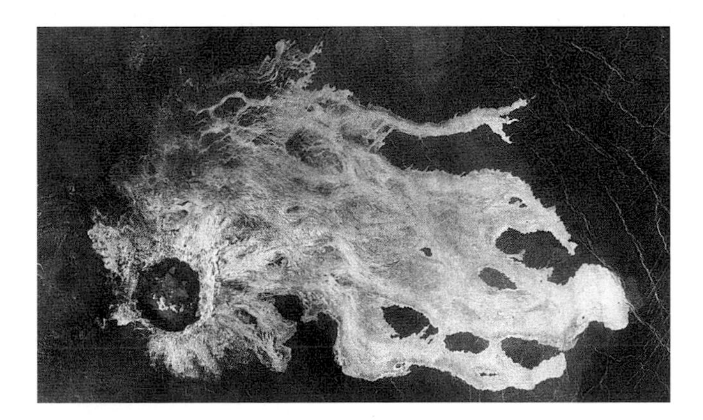

The peculiar effects of the dense Venusian atmosphere are manifested in these strange ejecta deposits, associated with the 72-km-diameter crater, Franklin. The flows appear to have travelled over 300 km and may represent large volumes of impact melt or fluidized ejecta.

have been significantly more plentiful and plate recycling might have been able to occur.

Assuming such a scenario, Ishtar Terra would be seen as a manifestation of subduction and collision processes that occurred before the present runaway greenhouse was in place. The implications of such an idea for Ishtar are that it must be a relatively ancient feature. This seems unlikely, and even if future research shows that it is a possibility, there remains the problem of the continued dynamical support needed to maintain its significant topography.

Long-term passive support can be ruled out – the high surface temperature of Venus means that failure of crustal rocks by ductile flow should be at lesser depths on Venus than on the Earth. Numerical models suggest that high-standing regions should spread under the influence of gravity by ductile flow of the relatively weak Venusian lower crust, on a time scale of the order of 10 million years. This implies that the geological processes which generate relief and steep slopes must have been active during the last 10 million years.

If major rises and coronae originate in similar mantle processes, which is widely accepted, then the implication is that mantle upwelling activity within Venus occurs on different scales. The large apparent depths of compensation associated with volcanic rises and the very extensive volcanism imply that they are connected with much larger-scale mantle upwellings than are coronae. Since most coronae are smaller than rises, it may be reasonably assumed that coronae represent a smaller scale of upwelling, or weaker upflow of material, or, perhaps, flows which are shorter-lived.

The primordial crust of Venus is probably not to be found anywhere on the planet's surface, and may have been rather like that of

The thick lava flow field seen in this image, of a part of Ovda Regio, shows that volcanism has also played a part in the geological development of highland massifs.

the lunar highlands. It is not known whether the Venusian crust that we do see is secondary, that is, predominantly basaltic, or tertiary, that is, reworked secondary material which might have a more evolved chemistry. The Venera 8 data suggest that some chemical differentiation has occurred, and this is supported by the now known occurrence of large steep-sided domes representing intrusions of more viscous magmas.

The complexity of the geology and the relative scarcity of impact craters makes stratigraphic studies difficult. All that can be said with any degree of certainty is that crater distribution studies suggest that resurfacing of the crust has been episodic and spatially inhomogeneous. Crater statistics appear to indicate that the mean age of the surface of the planet is between about 300 million to one billion years. The lower age may well pertain to the last major episode of resurfacing that Venus experienced. In this connection, it has been suggested by some planetary scientists that Venusian history may have been punctuated by major volcanic resurfacing episodes (enhanced heat flow) between which quiescent conditions prevailed while internal heat again built up. Further study of Magellan data should throw some light on this intriguing possibility.

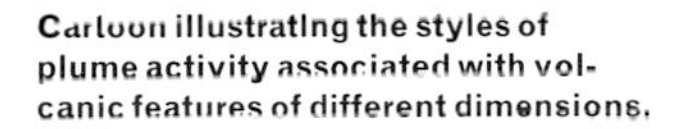

Cartoon illustrating the styles of plume activity associated with volcanic features of different dimensions.

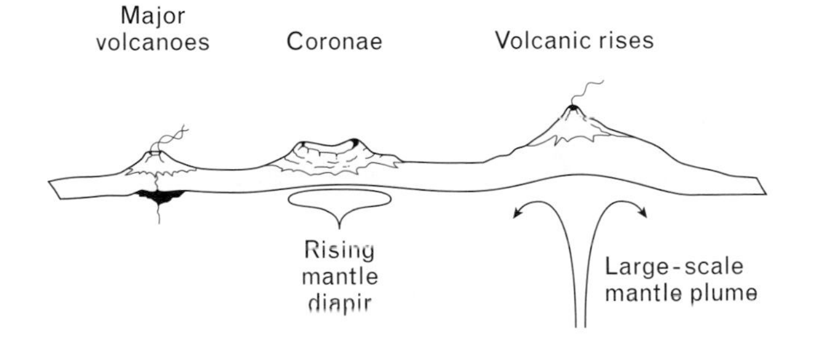

INTO THE FUTURE 12

From what has been said in preceding chapters, it is clear that Venus is a most unwelcoming world. Earth-type life there seems to be completely out of the question. Whether life ever appeared, at an early stage in Venus's history, is another matter; on the whole it appears rather improbable, and even if living things did gain a foothold they can hardly have persisted for long before conditions became overwhelmingly hostile.

We have learned more during the past few decades than we had been able to do throughout the whole of human history – after all, it was not so long ago that Venus was widely believed to be a watery world, and as a potential colony more promising than Mars! But much remains to be learned. No doubt there will be more missions during the coming century, and we may hope for sample-and-return probes, so that we can analyse the surface materials in our laboratories and search for any signs of past life; but the chances of any manned expeditions there seem to be nil. Of course, there have been suggestions that we may be able to "terraform" Venus, breaking down the atmospheric carbon dioxide and sulphuric acid and releasing free oxygen; but although this could be possible one day, it is quite beyond our present capabilities.

From the surface of Venus, the view would indeed be strange. There is no such thing as a "fine day"; the day is permanently overcast, and everything would be bathed in super-hot "smog", with the rocks reflecting the orange hue of the sky. Even the Sun would be forever hidden, and would betray its presence only by a local dull glare. A Venusian astronomer – if he or she could exist – could know little about the outer universe.

For the moment, and in the foreseeable future, we must therefore be content to study Venus from a respectful distance. Yet although it is not the friendly place we once believed it to be, it retains its fascination – and it is still the world closest to our own in size and mass, even though the Earth and Venus are non-identical twins.

APPENDICES

APPENDIX 1 SIGNIFICANT VENUS DATA

Parameter	Value
Mean distance from Sun	108,200,000 km
Sidereal period	224.701 days
Orbital inclination	3°.394
Orbital eccentricity	0.00678
Mean synodic period	583.92 days
Axial rotation period	243.01 days
Radius	6,051 ± 2 km
Mass (Earth = 1)	0.815
Volume (Earth = 1)	0.88
Mean surface pressure	90 bar
Mean density (water = 1)	5.25
Escape velocity	10.36 km s^{-1}
Mean visual opposition magnitude	–4.4
Albedo	0.76
Atmospheric main constituents	CO_2, N_2

APPENDIX 2 **SUMMARY OF VENUS SPACECRAFT MISSIONS**

(taken from *Greeley and Batson*, 1996)

Spacecraft	**Launch**	**Arrival**	**Remarks**
*Venera 1**	Feb. 12, 1961	May 19, 1961	Closest approach: 99,800 km; lost contact at 7 million km
Mariner 2	Aug. 26, 1962	Dec. 14, 1962	Flyby; closest approach: 34,830 km; first successful planetary flight
*Venera 2**	Nov. 12, 1965	Feb. 27, 1966	Closest approach: 23,950 km; failed to transmit data to Earth
*Venera 3**	Nov. 16, 1965	Mar. 1, 1966	Closest approach: impact; crushed by atmosphere at 32,000-m altitude; failed to return data
*Venera 4**	June 12, 1967	Oct. 18, 1967	Atmosphere probe; closest approach: impact (presumed); transmitted for 94 minutes during entry
Mariner 5	June 14, 1967	Oct. 19, 1967	Flyby; closest approach: 3,990 km
*Venera 5**	Jan. 5, 1969	May 16, 1969	Atmosphere probe; closest approach: impact (presumed); penetrated deeper than *Venera 4*
*Venera 6**	Jan. 10, 1969	May 17, 1969	Atmosphere probe; closest approach: impact (presumed); penetrated deeper than *Venera 4*
*Venera 7**	Aug. 17, 1970	Dec. 15, 1970	Lander; transmitted data 23 minutes from surface
*Venera 8**	Mar. 27, 1972	July 22, 1972	Lander; transmitted data 50 minutes from surface
Mariner 10	Nov. 3, 1973	Feb. 5, 1974	Flyby; closest approach: 5,310 km (*en route* to Mercury)
*Venera 9**	June 8, 1975	Oct. 23, 1975	Orbiter and lander; first photographs of surface
*Venera 10**	June 14, 1975	Oct. 26, 1975	Orbiter and lander; photographs of surface
Pioneer/Venus:			
orbiter	May 20, 1978	Dec. 4, 1978	Still operating as of Jan. 1992
multiprobe	Aug. 8, 1978	Dec. 9, 1978	Five atmosphere probes
*Venera 11**	Sept. 9, 1978	Dec. 21, 1978	Flyby and lander
*Venera 12**	Sept. 14, 1978	Dec. 25, 1978	Flyby and lander
*Venera 13**	Oct. 30, 1981	Mar. 1, 1982	Lander, surface data
*Venera 14**	Nov. 4, 1981	Mar. 5, 1982	Lander, surface data
*Venera 15**	June 2, 1983	Oct. 10, 1983	Orbiter, radar images of surface
*Venera 16**	June 7, 1983	Oct. 14, 1983	Orbiter, radar images of surface
[1]*Vega 1**	Dec. 15, 1984	June 11, 1985	Lander and balloon atmosphere probe
[1]*Vega 2**	Dec. 21, 1984	June 15, 1985	Lander and balloon atmosphere probe
Magellan	May 4, 1989	Aug. 10, 1990	Radar-imaging orbiter

* Soviet spacecraft; all others are/were from the United States except where noted

[1] The *Vega* main spacecraft both continued on to Comet Halley after dropping probes at Venus.

APPENDIX 3 PHENOMENA OF VENUS 1997–2005

Superior conjunctions

1997 April 2
1998 October 20
2000 June 11
2002 January 14
2003 August 18
2005 March 31

Inferior conjunctions

1998 January 16
1999 August 20
2002 October 31
2004 June 8

Greatest western elongations

1998 March 27
1999 October 30
2001 June 8
2003 January 11
2004 August 17

Greatest eastern elongations

1997 November 6
1999 June 11
2001 January 17
2002 August 22
2004 March 29
2005 November 3

On 2000 June 11, Venus will actually be occulted by the Sun, though of course the phenomenon will be completely unobservable!

The next transits will occur on 2004 June 8 (mid-transit 8h 24m UT), and 2012 June 6 (1h 36m UT). We must then wait until 2117 December 11 and 2125 December 8.

Occultations of bright stars or planets are very rare. On 1818 January 3, Venus occulted Jupiter, and will again do so on 2065 November 22. Venus occulted Regulus (α Leonis) on 1959 July 7, and Nunki (σ Sagittarii) on 1981 November 17.

APPENDIX 4 **ESTIMATED ROTATION PERIODS OF VENUS**

Well over a hundred estimates have been made; the following list includes some of the more notable attempts. The third column shows the method used: V = visual; S = spectroscopic; T = theoretical; P = photographic. The asterisk (*) indicates that the period was given as captured or synchronous (124d 16h 48m); an R indicates that the direction was thought to be retrograde. The estimates refer to the solid body of the planet, not the upper clouds, though until the modern era these two were thought (erroneously) to be the same thing.

Year	Observer	Method	Value
1666/7	G. D.Cassini	V	23h 21m
1727	F. Bianchini	V	24d 8h
1740	J. J. Cassini	V	23h 20m
1789	J. H. Schröter	V	23h 21m 19s
1811	J. H. Schröter	V	23h 21m 7s.977
1832	T. Hussey	T	24d 8h
1880	G. Schiaparelli	V	*
1881	W. F. Denning	V	23h 21m
1890	H. Perrotin	V	*
1890	F. Terby	V	Very slow or *
1892	V. Trouvelot	V	24h
1894	C. Flammarion	V	24h
1895	G. Schiaparelli	V	*
1895	A. S. Williams	V	24h
1896	L. Brenner	V	23h 57m 36s.27728
1897	H. McEwen	V	23h 30m
1900	A. Belopolsky	S	24h 42m
1903	V. M. Slipher	S	*
1909	P. Lowell	V,S	*
1911	A. Belopolsky	S	1d 11h
1915	W. Rabe	V	23h 57m
1919	W. Evershed	S	20h to 30h
1921	R. Jarry-Desloges	V	22h 53m
1921	W. H. Pickering	V	2d 10h
1922	A. Rordame	P	24h
1924	W. H. Steavenson	V	8d
1927	F. E. Ross	P	30h
1932	R. Barker	V	*
1933	E. P. Martz	V	1d 13h 4m 48s
1934	E. M. Antoniadi	V	Very long or *
1939	W. H. Haas	V	*
1949	V. V. Volkov	V	2d 12h
1951	R. M. Baum	V	195d
1953	G. D. Roth	V	15h
1954	G. P. Kuiper	P	A few weeks
1955	A. Dollfus	V,P	*
1956	J. D. Kraus	Radio	22h 17m
1958	R. S. Richardson	S	Slow, R?
1962	R. L. Carpenter	Radar	c. 250d, R
1962	R. M. Goldstein	Radar	248d ± 50d, R
1962	F. D. Drake	Radar	slow, R
1964	I. I. Shapiro	Radar	247d±4d R
1964	R. L. Carpenter	Radar	258d ± 6d, R
1979	I. I. Shapiro	Radar	243d.01 ± 0d.03

The modern value is 243d.16. The "solar day" on Venus is 117 Earth-days; that is to say, a daylight period of 58.5 days followed by a night of equal length.

APPENDIX 5 **NAMES OF VENUSIAN FEATURES**

Adapted from *The NASA Atlas of the Solar System*, by Ronald Greeley and Raymond Batson (Cambridge 1997), and used courtesy of the United States Geological Survey, Flagstaff, Arizona

[a] Note: 'Unqualified' names are all CRATERS. Other features are qualified by type (e.g. Corona)

Feature name[a]	Lat. (°)	Long. (°)	Size (km)	Description
Abaka	52.6S	104.2E	14	Maori first name.
Abigail	52.2S	111.1E	19	Hebrew first name.
Abington	47.8S	277.8E	23	Frances; English actress (1737-1815).
Abundia Corona	18.5N	125.0E	250	Norse goddess of giving.
Adaiah	47.3S	253.3E	19	Hebrew first name.
Adamson	14.8S	29.6E	20	Joy; Austrian-born British author, animal expert (1910-1980).
Addams	56.1S	98.9E	90	Jane; American social reformer (1860-1935).
Adivar	8.9N	75.9E	30	Halide; Turkish educator, author (1883-1964).
Aegina Farrum	35.5N	20.9E	60	Greek river nymph.
Aeracura Corona	19.0S	238.5E	250	Celtic earth goddess.
Aethelflaed	18.2S	196.5E	17	English leader of the Mercians (ca. 884-918).
Aglaonice	26.5S	339.9E	66	Ancient Greek astronomer.
Agnesi	39.5S	37.8E	40	Maria; Italian mathemetician (1718-1799).
Agraulos Corona	27.7S	165.8E	170	Greek fertility goddess.
Agrippina	33.2S	65.2E	37	Roman empress (ca. 13 B.C.-A.D. 33).
Ahsonnutli Dorsa	47.9N	194.8E	1708	Navajo (N. America) spirit of light and sky.
Aimée	16.2N	127.1E	18	French first name.
Aino Planitia	40.5S	94.5E	4983	Finnish heroine who became water spirit.
Aita	8.9N	270.7E	14	Estonian first name.
Aitchison Patera	16.7S	349.4E	28	Alison; American geographer.
Akeley	8.0N	244.4E	25	Delia; American explorer (1875-1970).
Akhmatova	61.1N	307.4E	42	Anna; Russian poet (1889-1966).
Akiko	30.7N	187.1E	24	Yosano; Japanese Tanka poetess (1878-1942).
Akkruva Colles	46.1N	115.5E	1059	Saami-Lapp fishing goddess.
Akna Montes	68.9N	318.2E	830	Mayan goddess of birth.
Aksentyeva	42.0S	271.9E	43	Zinaida; Soviet geophysicist, astronomer (1900-1969).
Alcott	59.5S	354.5E	71	Louisa M.; American author (1832-1888).
Ale Tholus	68.2N	247.0E	87	Igbo (Nigeria) goddess who created Earth and vegetation.
Alima	45.9S	229.2E	13	Tatar first name.
Alimat	29.6S	205.9E	14	Osset first name.
Alison	4.0S	165.6E	16	Irish first name.
Allat Dorsa	63.3N	71.3E	302	Arab sky goddess.
Allatu Corona	15.5N	114.0E	125	Akkadian earth goddess.
Alma	2.4S	228.7E	17	Kazakh first name.

Feature name[a]	Lat. (°)	Long. (°)	Size (km)	Description
Almeida	46.0N	123.0E	18	Portuguese first name.
Alpha Regio	25.5S	1.3E	1897	First letter in Greek alphabet.
Al-Taymuriyya	32.9N	336.2E	22	Ayesha; Egyptian author, feminist (1840-1902).
Al-Uzza Undae	67.7N	90.5E	150	Arabian desert goddess.
Amalasthuna	11.5S	342.4E	18	Ostrogoth queen (ca. 498-535).
Amaya	11.3N	89.1E	32	Carmen; Spanish Gypsy dancer (1913-1963).
Amenardes	15.0N	54.1E	25	Egyptian princess (718-655 B.C.).
Ament Corona	67.2S	217.9E	115	Egyptian earth goddess.
Anahit Corona	77.1N	277.3E	324	Armenian goddess of fertility.
Anala Corona	11.0N	14.0E	240	Hindu fertility goddess.
Ananke Tessera	53.3N	133.3E	1060	Greek goddess of necessity.
Anaxandra	44.2N	162.1E	21	Greek artist (flourished ca. 228 B.C.).
Andami	17.5S	26.3E	28	Iranian doctor.
Andreianova	3.0S	68.7E	70	Elena; Russian ballerina (ca. 1821-ca. 1855).
Anicia	26.4S	31.1E	30	Greek physician, poet (flourished ca. 300 B.C.).
Annapurna Corona	35.5S	152.0E	300	Indian goddess of wealth.
Annia Faustina	22.1N	4.6E	20	Roman empress, wife of Marcus Aurelius (125-175).
Anning Paterae	66.5N	57.8E	—	Mary; English paleontologist (1799-1847).
Anqet Farrum	33.6N	311.5E	125	Egyptian goddess of fertile waters.
Anthony Patera	48.0N	33.0E	—	Susan B.; American suffrage leader (1820-1906).
Antiope Linea	40.0S	350.0E	—	Amazon awarded to Theseus.
Antonina	27.0N	107.1E	18	Russian first name.
Anuket Vallis	66.7N	8.0E	350	Egyptian river goddess.
Anush	14.9N	86.5E	13	Armenian first name.
Anya	39.5N	298.2E	21	Russian first name.
Apgar Patera	43.5N	84.0E	—	Virginia; American doctor (1909-1974).
Aphrodite Terra	5.8S	104.8E	9990	Greek goddess of love.
Api Mons	38.9N	54.7E	190	Scythian goddess of earth.
Aramaiti Corona	26.3S	82.0E	350	Persian fertility goddess.
Aranyani Chasma	69.3N	74.4E	718	Indian forest goddess.
Ariadne	43.8N	180.0E	37	Greek first name; crater defines longitude.
Arianrod Fossae	37.0N	239.9E	715	Celtic warrior queen.
Artemis Chasma	41.2S	138.5E	3087	Greek goddess of hunt/moon.
Artemis Corona	35.0S	135.0E	2600	Named from associated chasma.
Aruru Corona	9.0N	262.0E	450	Sumerian earth goddess.
Ashnan Corona	50.2N	357.0E	300	Sumerian harvest goddess.
Ashtart Tholus	48.7N	247.0E	138	Phoenician goddess of love, fertility and war.
Asmik	3.9N	166.4E	18	Armenian first name.
Aspasia Patera	56.4N	189.1E	150	Outstanding woman of ancient Greece (ca. 470-410 B.C.).
Asteria Regio	21.6N	267.5E	1131	Greek Titaness.

Feature name[a]	Lat. (°)	Long. (°)	Size (km)	Description
Astrid	21.4S	335.5E	12	Scandanavian first name.
Atalanta Planitia	45.6N	165.8E	2048	Greek; huntress associated with golden apples.
Atargatis Corona	8.0S	8.6E	360	Hittite fertility goddess.
Atete Corona	16.0S	243.5E	600	Oromo (Ethiopia) fertility goddess.
Atira Mons	52.2N	267.6E	152	Pawnee (N. America) wife of Great Spirit Tirawa.
Atla Regio	9.2N	200.1E	3200	Norse giantess, mother of Heimdall.
Atropos Tessera	71.5N	304.0E	469	Greek; one of three Fates.
Atse Estsan Corona	8.5N	92.0E	150	Navajo fertility goddess.
Audhumla Corona	45.5N	12.0E	225	Norse primordial nourisher.
Audra Planitia	61.5N	71.5E	—	Lithuanian sea mistress.
Audrey	23.8N	348.1E	15	English first name.
Aurelia	20.3N	331.8E	31	Mother of Julius Caesar.
Auska Dorsum	60.5N	179.0E	—	Lithuanian goddess of sun rays.
Aušrā Dorsa	49.4N	25.3E	859	Lithuanian dawn goddess.
Austen	25.0S	168.3E	46	Jane; English novelist (1775-1817).
Avfruvva Vallis	2.0N	70.0E	70	Saami (Lapp) river goddess.
Avviyar	18.0S	353.6E	21	Tamil poet (ca. 100 B.C.).
Ayana	29.1S	175.5E	15	Altai first name.
Ayrton Patera	6.0N	228.3E	85	Hertha M.; English physicist (1854-1923).
Baba-Jaga Chasma	53.2N	49.5E	580	Slavic forest witch.
Bachue Corona	73.3N	261.4E	463	Chibcha (Colombia) goddess of fertility.
Badarzewska	22.6S	137.0E	28	Thekla; Polish composer (1834-1861).
Ba'het Patera	48.6N	0.6E	—	Egyptian who defeated Portugese.
Baker	62.6N	40.5E	105	Josephine; American expatriate dancer, singer (1906-1975).
Balch	29.9N	282.9E	37	Emily; American economist, Nobel laureate (1867-1961).
Baltis Vallis	37.3N	161.4E	6000	Syrian word for planet Venus.
Ban Zhao	17.2N	146.9E	38	Chinese historian (ca. 35-100).
Baranamtarra	17.9N	267.8E	25	Mesopotamian queen (ca. 2500 B.C.).
Barrera	16.6N	109.3E	25	Olivia; Spanish medical writer (born 1562).
Barrymore	52.3S	195.6E	50	Ethel; American actress (1879-1959).
Barsova	61.5N	223.3E	95	Valeria; Soviet singer (1892-1967).
Barto	45.2N	146.4E	54	Agniya; Soviet poet (1906-1981).
Barton	27.4N	337.5E	50	Clara; American Red Cross founder (1821-1912).
Bascom	10.3S	302.2E	36	Florence; American geologist (1862-1945).
Bashkirtseva	14.7N	194.0E	38	Marie; Russian painter, diarist (ca. 1859-1884).
Bassi	19.0S	64.6E	35	Laura; Italian physicist, mathematician (1711-1778).
Bast Tholus	57.8N	130.3E	83	Egyptian goddess of joy.
Bathsheba	15.1S	49.3E	36	Hebrew queen (ca. 1030 B.C.).
Bau Corona	53.0N	258.0E	—	Sumerian fertility goddess.
Bayara Vallis	45.6N	16.5E	500	Dogon (Mali) word for planet Venus.
Bécuma Mons	34.0N	22.0E	—	Irish goddess.

Feature name[a]	Lat. (°)	Long. (°)	Size (km)	Description
Beecher	13.1N	253.5E	35	Catherine; American educator, author (1800-1878).
Behn	32.5S	141.8E	25	Aphra; English novelist, poet, playwright (1640-1689).
Beiwe Corona	52.6N	306.5E	600	Saami (Lapp) fertility goddess.
Belet-Ili Corona	6.0N	20.0E	300	Mesopotamian nature/fertility goddess.
Belisama Vallis	50.0N	22.5E	220	English Celtic river goddess.
Bell Regio	32.8N	51.4E	1778	English giantess.
Bellona Fossae	38.0N	222.1E	855	Roman war goddess, wife of Mars.
Ben Dorsa	71.2N	284.1E	628	Vietnamese sky goddess.
Bennu Vallis	1.3N	341.2E	710	Egyptian word for planet Venus.
Benten Corona	16.0N	340.0E	310	Japanese love/fertility goddess.
Bereghinya Planitia	28.6N	23.6E	3902	Slavic water spirit.
Berggolts	63.4S	53.0E	31	Olga; Russian poet (1910-1975).
Bernadette	46.7S	285.6E	15	French first name.
Bernhardt	31.4N	84.3E	45	Sarah; French actress (1844-1923).
Berta	62.0N	322.0E	20	Finnish first name.
Beruth Corona	19.0S	233.5E	350	Phoenician earth goddess.
Beta Regio	25.3N	282.8E	2869	Second letter in Greek alphabet.
Bethune Patera	47.0N	321.5E	—	Mary; American educator (1875-1955).
Bette	24.6S	347.9E	6	German first name (form of Elizabeth).
Beyla Corona	25.0N	15.5E	400	Norse earth goddess.
Bezlea Dorsa	30.4N	36.5E	807	Lithuanian evening light goddess.
Bhumidevi Corona	17.2S	343.6E	150	Hindu earth goddess.
Bhumiya Corona	15.0N	118.0E	100	Hindu earth goddess.
Bickerdyke	82.0S	170.8E	39	Mary; American Civil War nurse (1817-1901).
Birute	36.0N	33.5E	—	Lithuanian first name.
Blackburne	11.0N	183.8E	33	Anna; English biologist (1726-1793).
Blai Corona	0.4S	134.5E	125	Celtic fertility goddess.
Blanche	8.5S	157.8E	18	French first name.
Blathnat Corona	35.0N	293.8E	300	Celtic fertility goddess.
Blixen	59.9S	145.6E	22	Karen; Danish author (1885-1962).
Bly	37.7N	305.5E	20	Nellie; American journalist (1867-1892).
Boadicea Paterae	56.0N	96.0E	—	(Boudicca); queen and heroine of Iceni (English Celtic tribe).
Boann Corona	27.0N	136.5E	300	Irish fertility goddess.
Boivin	4.3N	299.5E	18	Marie; French medical researcher (1773-1847).
Boleyn	24.4N	219.9E	69	Anne; English queen (1507-1536).
Bona Corona	24.0S	157.5E	275	Roman virgin/fertility goddess.
Bonnevie	36.1S	126.8E	85	Norwegian biologist.
Bonnin	6.2S	117.6E	40	Gertrude (Zitkala-sa); Dakota reformer, writer (1875-1938).
Boulanger	26.5S	99.3E	57	Nadia; French pianist, composer (1887-1979).
Bourke-White	21.2N	147.8E	31	Margaret; American photo-journalist (1906-1971).
Boyd	39.3S	221.3E	25	Louise; American explorer (1887-1972).

Feature name[a]	Lat. (°)	Long. (°)	Size (km)	Description
Boye	9.6S	292.3E	30	Karin; Swedish poet, novelist (1900-1941).
Bradstreet	16.5N	47.6E	39	Anne; American poet (ca. 1612-1672).
Breksta Dorsa	35.9N	304.0E	700	Lithuanian night darkness goddess.
Bremer Patera	67.0N	64.0E	—	Frederika; Swedish writer, reformer, feminist (1801-1865).
Bridgit	45.3S	348.9E	11	Irish first name.
Brigit Tholus	49.0N	246.0E	—	Celtic goddess of wisdom, doctoring, smithing.
Britomartis Chasma	33.0S	130.0E	0	Greek/Cretan goddess of the hunt.
Brooke	48.0N	296.0E	26	Frances; Canadian novelist (1724-1789).
Browning	28.0N	5.0E	24	Elizabeth; British poet (1806-1861).
Bryce	62.6S	197.1E	25	Lucy; Australian medical pioneer (1897-1968).
Buck	5.7S	349.6E	22	Pearl S.; American writer (1892-1973).
Budevska	0.5N	143.0E	20	Adriana; Bulgarian actress (1878-1955).
Bugoslavskaya	23.0S	300.4E	30	Yevgenia; Soviet astronomer (1899-1960).
Bunzi Mons	46.0N	355.0E	—	Woyo (Zaire) rainbow goddess.
Caccini	17.4N	170.4E	38	Francesca; Italian poet, composer (ca. 1581-ca. 1640).
Cailleach Corona	48.0S	88.3E	125	Scottish Celtic fertility goddess.
Caitlin	65.3S	12.1E	14	Welsh first name.
Caiwenji	12.4S	287.5E	22	Chinese painter, calligrapher (907-960).
Calakomana Corona	6.5N	43.5E	575	Pueblo Indian corn goddess.
Caldwell	23.6N	112.1E	44	Taylor; American author (1900-1985).
Callas	2.4N	26.9E	30	Maria; American opera singer (1923-1977).
Callirhoe	21.3N	140.6E	32	Greek sculptor (ca. 600 B.C.).
Carmenta Farra	12.4N	8.0E	180	Roman goddess of springs.
Caroline	6.8N	306.3E	17	First name from French.
Carpo Corona	37.5S	3.0E	215	Greek fertility goddess.
Carr	24.0S	295.7E	30	Emily; Canadian artist (1871-1945).
Carreño	3.9S	16.1E	57	Teresa; Venezuelan pianist, composer (1853-1917).
Carriera Patera	48.5N	48.5E	—	Rosalba; Italian portrait painter (1675-1757).
Carson	24.2S	344.2E	41	Rachel; American biologist, author (1907-1964).
Carter	5.3N	67.2E	18	Maybelle; American singer, songwriter (1909-1978).
Cassatt Patera	65.5N	207.5E	—	Mary; American Impressionist painter (1845-1926).
Castro	3.3N	233.9E	23	Rosalie; Spanish poet, novelist (1837-1885).
Cather	47.1N	106.7E	30	Willa; American novelist (1876-1947).
Cauteovan Corona	31.5N	144.0E	—	Kataba (Colombia) fertility goddess.
Cavell Patera	38.0N	19.0E	—	Edith; British nurse, heroine (1865-1915).
Centlivre	19.1N	290.4E	26	Susannah; English actress, playwright (ca. 1667-1723).
Ceres Corona	16.0S	151.5E	675	Roman harvest goddess.
Chapelle	6.4N	103.8E	23	Georgette; Am. photo-journalist, killed in Viet Nam (1919-1965).
Chih Nu Dorsum	73.0S	195.0E	625	Chinese sky goddess.
Chiun Corona	18.3N	340.5E	150	Hebrew fertility goddess.
Chiyojo	47.8S	95.2E	35	Japanese poetess.

Feature name[a]	Lat. (°)	Long. (°)	Size (km)	Description
Chloe	7.4S	98.6E	19	First name from Greek.
Christie	28.1N	72.5E	34	Agatha; British novelist (1890-1976).
Citlalpul Valles	57.4S	185.0E	2350	Aztec name for planet Venus.
Ciuacoatl Corona	53.0N	150.9E	100	Aztec earth goddess.
Cleopatra	66.0N	8.0E	104	Egyptian queen, (69-30 B.C.).
Cline	21.8S	317.0E	40	Patsy; American singer (1932-1963).
Clotho Tessera	56.4N	334.9E	289	Greek; one of three Fates.
Coatlicue Corona	63.2N	273.0E	199	Aztec earth goddess.
Cochran	52.0N	142.6E	124	Jacqueline; American aviator (ca. 1910-1980).
Cohn	33.2S	208.1E	21	Carola; Australian artist (1892-1964).
Colette Patera	66.5N	322.8E	149	Sidonie Gabrielle; French novelist (1873-1954).
Colijnsplaat Corona	32.0S	151.0E	350	Teutonic fertility goddess.
Colleen	60.8S	162.2E	14	Irish first name.
Colonna	64.7N	216.8E	28	Vittoria; Italian poet (ca. 1490-1547).
Comnena	1.2N	343.7E	20	Anna; Byzantine princess, physician, writer (1083-1148).
Conway	48.3N	39.1E	50	Anne Finch; English natural scientist (1631-1679).
Copia Corona	42.5S	75.5E	500	Roman goddess of plenty.
Corday Patera	62.5N	40.0E	—	Charlotte; French patriot (1768-1793).
Cori	25.4N	72.7E	50	Gerty; Czech biochemist, Nobel laureate (1896-1957).
Corinna	22.8N	40.5E	21	Greek poet (flourished ca. 490 B.C.).
Corpman	0.3N	151.8E	52	Elizabeth; Polish astronomer, wife of Hevelius (17th century).
Cortese	11.4S	218.3E	28	Isabella; Italian physician, medical writer (died 1561).
Cotis Mons	44.1N	233.1E	62	Thracian goddess, mother of gods, similar to Cybele.
Cotton	71.0N	300.0E	40	Egenni; French physicist (1881-1967).
Cunitz	14.5N	350.9E	48	Maria; Polish astronomer-mathematician (1610-1664).
Cybele Corona	7.5S	20.7E	500	Phrygian fertility goddess.
Cynthia	16.7S	347.5E	19	First name from Greek.
Dali Chasma	17.6S	167.0E	2077	Georgian; goddess of hunt.
Danilova	26.4S	337.3E	50	Maria; Russian ballet dancer (born 1793).
Danu Montes	58.5N	334.0E	808	Celtic mother of gods.
Danute	63.5S	56.5E	14	Lithuanian first name.
Daphne	41.3N	280.4E	14	First name from Greek.
Darago Fluctus	11.5S	313.5E	775	Philippine volcano goddess.
Darline	19.3S	232.6E	13	Anglo-Saxon first name.
Dashkova	77.9N	305.7E	42	Yekaterina; Russian philologist (1743-1810).
Datsolalee	38.3N	171.6E	19	Washo Indian artist, basketmaker (1835-1925).
Daura Chasma	72.4N	53.8E	729	Hausa (W. Sudan) great huntress.
Davies Patera	47.0N	269.0E	—	Sarah Emily; British educator; college founder (1830-1921).
de Ayala	12.3N	31.9E	20	Josefa; Spanish painter (1630-1684).
de Beausoleil	5.0S	102.9E	30	Martine; French earth science researcher (17th century).
de Beauvoir	2.0N	96.0E	40	Simone; French writer (1908-1986).

Feature name[a]	Lat. (°)	Long. (°)	Size (km)	Description
de Lalande	20.3N	354.9E	20	Marie-Jeanne; French astronomer (1768-1832).
de Staël	37.4N	324.2E	25	Anne; French historian, novelist (1766-1817).
De Witt	6.5S	275.6E	21	Lydia; American pathologist (1859-1928).
Deken	47.0N	288.5E	58	Agatha; Dutch novelist (1741-1804).
Dekla Tessera	57.4N	71.8E	1363	Latvian goddess of fate.
Deledda	76.0N	127.5E	32	Grazia; Italian novelist (1871-1936).
Delilah	57.9S	250.5E	18	First name from Hebrew.
Deloria	32.0S	97.0E	38	Ella; Dakota (Sioux) anthropologist (1888-1971).
Demeter Corona	53.9N	294.8E	560	Greek goddess of fertility.
Dennitsa Dorsa	85.6N	205.9E	872	Slavic goddess of day, light.
Derceto Corona	46.8S	20.2E	200	Phillistine fertility goddess.
d'Este	34.2S	238.7E	21	Isabella; Italian archaeologist, businesswoman (1474-1539).
Devana Chasma	9.6N	284.4E	1616	Czechoslovakian goddess of hunt.
Devorah	22.5S	343.4E	6	Hebrew first name.
Devorguilla	15.3N	3.8E	22	Irish heroine (died 1193).
Dhisana Corona	14.5N	111.7E	100	Vedic goddess of plenty.
Diana Chasma	14.8S	154.8E	938	Roman goddess of hunt/moon.
Dickinson	74.7N	177.2E	54	Emily; American poet (1830-1886).
Dinah	62.8S	37.0E	19	Hebrew first name.
Dione Regio	31.5S	328.0E	2300	Greek Titaness; 1st wife of Zeus; mother of Aphrodite (Venus).
Dix	36.9S	329.1E	68	Dorothea; American nurse, reformer (1802-1887).
Dodola Dorsa	46.8N	272.6E	607	South Slavic rain goddess.
Dolores	51.5N	200.5E	16	Spanish first name.
Doris	2.3N	89.9E	16	First name from Greek.
Drena	20.6S	338.6E	2	Lithuanian first name.
du Chatelet	21.5N	165.0E	19	Gabrielle Emilie; French mathematician, physicist (1706-1749).
Duncan	67.9N	291.7E	38	Isadora; American dancer (1878-1927).
Durant	62.3S	227.5E	23	Ariel; American writer (1898-1981).
Duse	82.5S	358.0E	27	Eleonora; Italian actress (1859-1924).
Dyan-Mu Dorsa	78.2N	31.9E	687	Chinese lightning goddess.
Earhart Corona	70.1N	136.2E	414	Amelia; American aviatrix (1897-1937).
Edgeworth	32.0N	22.7E	35	Maria; British novelist (1767-1849).
Edinger	68.8S	208.3E	34	Tilly; American geologist (1897-1967).
Efimova	81.0N	224.0E	28	(Simonovich-Efimova) Nina; Soviet painter and puppet-theatre designer (1877-1948).
Egeria Farrum	43.6N	7.5E	40	Roman water nymph.
Eigin Corona	5.0S	175.0E	200	Celtic fertility goddess.
Eileen	22.8S	232.6E	15	Irish first name.
Eistla Regio	10.5N	21.5E	8015	Norse giantess.
Eithinoha Corona	57.0S	7.5E	500	Iroquois earth goddess.
Elena	18.3S	73.3E	18	Italian first name.
Eliot Patera	39.0N	79.0E	—	George (Mary Ann Evans); English writer (1819-1880).

Feature name[a]	Lat. (°)	Long. (°)	Size (km)	Description
Elza	34.4S	275.8E	17	Latvian first name.
Enyo Fossae	61.0S	344.0E	900	Greek war goddess.
Eostre Mons	45.0N	329.5E	—	Teutonic goddess of spring.
Epona Corona	28.0S	208.5E	225	Celtic horse/fertility goddess.
Ereshkigal Corona	21.0N	84.5E	320	Mesopotamian nature/fertility goddess.
Erika	72.0N	176.0E	12	Hungarian, German first name.
Erin	47.0S	184.8E	14	Irish first name.
Erinna	78.0S	309.0E	33	Greek poet (7th century B.C.).
Eriu Fluctus	35.0S	358.0E	1200	Irish earth mother.
Erkir Corona	16.3S	233.7E	275	Armenian earth goddess.
Ermolova	60.0N	154.0E	60	Mariya; Russian actress (1853-1928).
Erxleben	50.9S	39.3E	28	Dorothea; first woman Ph.D. in Germany (1715-1762).
Escoda	18.2N	149.4E	20	Philipino organizer of Girl Scouts of the Philippines (1898-1945).
Estelle	1.1N	93.7E	20	First name from Latin.
Esther	19.4N	21.8E	17	First name from Persian.
Eudocia	59.5S	201.8E	28	Byzantine empress (ca. 401-460).
Eurynome Corona	26.5N	94.5E	200	Greek mother earth goddess.
Evangeline	69.7N	221.8E	15	First name from Greek.
Eve Corona	32.0S	359.8E	330	Hebrew first name; name changed from Eve (crater).
Evika	5.1S	31.4E	16	Tatar first name.
Faiga	4.9N	170.9E	10	Anglo-Saxon first name.
Fakahotu Corona	60.3N	100.4E	290	Tuamotu earth mother.
Farida	4.8N	38.9E	20	Azerbaijan first name.
Fatima	17.8S	31.9E	15	Arabic first name.
Fatua Corona	16.3S	17.7E	400	Roman goddess of fertility.
Fedorets	59.8N	65.7E	45	Velentina; Soviet astronomer (1923-1976).
Fedosova	45.0N	171.8E	24	Irina; Russian folk poet (1831-1899).
Felesta Fossae	35.0N	46.5E	—	Amazon queen in Scythian epic tales.
Felicia	19.8S	226.4E	12	First name from Latin.
Ferber	26.4N	13.0E	23	Edna; American author (1887-1968).
Fernandez	76.3N	16.4E	26	M. A.; Spanish actress (18th century).
Feronia Corona	68.0N	281.7E	360	Ancient Italian goddess of spring and flowers.
Ferrier	15.8N	111.1E	30	Kathleen; English opera singer (1912-1953).
Festa	11.5N	27.2E	25	Italian painter.
Flagstad	54.3S	18.9E	48	Kirsten; Norwegian opera singer (1895-1962).
Flosshilde Farra	10.5N	279.4E	75	German water nymph.
Foquet	15.1S	203.5E	50	Marie; French medical writer, charity worker (17th century).
Fornax Rupes	30.3N	201.1E	729	Roman goddess of hearth and baking of bread.
Fortuna Tessera	69.9N	45.1E	2801	Roman goddess of chance.
Fossey	2.0N	188.7E	30	Diane; American zoologist, conservationist (1932-1985).
Fotla Corona	58.5S	163.5E	150	Celtic fertility goddess.

Feature name[a]	Lat. (°)	Long. (°)	Size (km)	Description
Francesca	28.0S	57.7E	18	Italian first name.
Frank	13.2S	12.9E	20	Anne; Dutch heroine, diarist (1929-1945).
Fredegond	50.7S	92.9E	26	Frankish queen (died A.D. 597).
Freyja Montes	74.1N	333.8E	579	Norse, mother of Odin.
Friagabi Fossae	50.2N	109.5E	141	Old English goddess, connected with Mars.
Frida	68.0N	56.0E	24	Swedish first name.
Frigg Dorsa	51.2N	148.9E	896	Norse, wife of supreme god Odin.
Fukiko	23.2S	105.7E	15	Japanese first name.
Furki Mons	35.9N	236.4E	79	Chechen and Ingush (Caucasus) goddess, wife of thunder god Sela.
Gabie Rupes	67.5N	109.9E	350	Lithuanian goddess of fire and hearth.
Gabriela	17.9S	240.4E	19	First name from Hebrew.
Gaia	6.0N	21.5E	400	Greek earth/fertility goddess.
Galina	47.4N	307.0E	20	Bulgarian first name.
Galindo	23.3S	258.8E	24	Beatrix; Italian physician, educator (1473-1535).
Ganiki Planitia	25.9N	189.7E	5158	Orochian (Siberia) water spirit, mermaid.
Ganis Chasma	16.3N	196.4E	615	Western Lapp forest maiden.
Gautier	26.5N	42.8E	60	Judith; French novelist (1845-1917).
Gaze	17.9N	240.2E	30	Vera; Soviet astronomer (1899-1954).
Gefjun Corona	33.5S	98.5E	300	Norse fertility goddess.
Gentileschi	45.2N	260.5E	20	Artemisia; Italian painter (1593-ca. 1652).
Gerda	45.9N	91.0E	30	Danish, German first name.
Germain	38.0S	63.5E	33	Sophie; French mathematician (1776-1831).
Gertjon Corona	30.0S	276.0E	250	Teutonic goddess of fertility.
Giliani	72.9S	142.0E	27	Alessandra; Italian anatomist (1307-1326).
Gillian	15.2S	49.9E	16	First name from Latin.
Gilmore	6.6S	132.8E	23	Mary; Australian poet (1865-1962).
Gina	78.0N	76.0E	24	Italian first name.
Glaspell	58.4S	269.6E	26	Susan; American playwright, novelist (ca. 1882-1948).
Gloria	68.5N	94.5E	14	Portuguese first name.
Godiva	56.1S	251.5E	32	(Godgifu); Mercian (England) noblewoman (ca. 1040-1085).
Golubkina	60.0N	286.5E	27	Anna; Soviet sculptor (1864-1927).
Goncharova	63.0S	97.7E	30	Natalya; Russian artist (1881-1962).
Goppert-Mayer	59.8N	26.5E	35	Maria; Polish physicist, Nobel laureate (1907-1972).
Grace	13.9S	268.9E	19	First name from Greek.
Graham	6.0S	6.0E	75	Martha; American dancer, choreographer (1894-1991).
Grazina	72.5N	337.5E	16	Lithuanian first name.
Greenaway	22.9N	145.1E	85	Catherine (Kate); English author, illustrator (1846-1901).
Gregory	7.2N	95.8E	21	Isabella; Irish playwright (1852-1932).
Gretchen	59.6S	213.2E	20	German first name.
Grey	52.4S	329.2E	50	Jane; English noblewoman (1537-1554).
Grimké	17.3N	215.3E	37	Sarah; American abolitionist (1792-1873).

Feature name[a]	Lat. (°)	Long. (°)	Size (km)	Description
Guan Daosheng	61.1S	181.8E	46	Chinese painter, calligrapher (1262-1319).
Gudrun	10.6N	326.3E	15	First name from Norse.
Guilbert	57.9S	13.3E	30	Yvette; French singer (1869-1944).
Guinevere Planitia	21.9N	325.0E	7519	British, wife of Arthur.
Gula Mons	21.9N	359.1E	276	Babylonian earth mother, creative force.
Guor Linea	17.0N	2.6E	—	Northern European Valkyrie.
Gwynn	9.7N	37.2E	32	Nell; English actress, courtesan (1650-1687).
Habonde Corona	3.0N	81.8E	125	Danish goddess of abundance.
Hallé	19.8S	145.4E	23	Wilhelmina; Austrian violinist (1839-1911).
Hallgerda Mons	55.0N	198.0E	—	Icelandic goddess of vanity.
Hannah	17.9N	102.6E	19	First name from Hebrew.
Hansberry	22.7S	324.1E	28	Lorraine; American playwright (1930-1965).
Hariasa Linea	19.0N	15.0E	—	German war goddess.
Hathor Mons	38.7S	324.7E	333	Egyptian sky goddess.
Hatshepsut Patera	28.1N	64.5E	118	Egyptian queen (1479 B.C.); pretended to be male.
Haumea Corona	54.0N	21.8E	375	Polynesian fertility goddess.
Hayasi	53.7N	244.1E	38	Fumiko; Japanese writer (1903-1951).
Heather	6.7S	334.1E	12	English first name.
Hecate Chasma	18.2N	254.3E	3145	Greek moon goddess.
Heidi	23.6N	350.1E	14	First name; form of Hester.
Helen Planitia	51.7S	263.9E	4362	Greek; "the face that launched 1000 ships."
Hellman	4.8N	356.2E	24	Lillian; American playwright, author (1907-1984).
Héloise	40.0N	51.9E	40	French physician, hospital founder (ca. 1098-1164).
Hemera Dorsa	51.0N	243.4E	587	Greek goddess, personification of day.
Heng-o Chasma	6.6N	355.5E	734	Chinese moon goddess.
Heng-o Corona	2.0N	355.0E	1060	Named for associated chasma.
Henie	52.0S	145.8E	70	Sonja; Norwegian skater (1912-1969).
Henwen Fluctus	20.5S	179.9E	485	British Celtic sow-goddess.
Hepat Corona	2.0S	145.5E	150	Hittite mother goddess.
Hepworth	5.2N	94.7E	54	Barbara; English sculptor (1903-1975).
Hera Dorsa	36.4N	29.5E	813	Greek sky goddess, wife of Zeus.
Hervor Corona	25.5S	269.0E	250	Norse fertility goddess.
Hestia Rupes	6.0N	71.1E	588	Greek hearth goddess.
Hiei Chu Patera	48.3N	97.4E	139	Chinese, converted silk worm product into thread (2698 B.C.).
Higgins	7.6N	241.4E	40	Marguerite; American journalist (1920-1966).
Hildr Fossa	45.4N	159.4E	677	Norse mythological warrior.
Himiko	19.0N	124.2E	35	Japanese queen (4th century A.D.).
Hina Chasma	64.5N	20.0E	—	Hawaiian moon goddess.
Hippolyta Linea	42.0S	345.0E	—	Amazon queen.
Holde Corona	53.5N	155.8E	200	German fertility goddess.
Holiday	46.7S	12.7E	24	Billie; American singer (1915-1959).

Feature name[a]	Lat. (°)	Long. (°)	Size (km)	Description
Horner	23.4N	97.5E	28	Mary; 19th century English naturalist, geologist.
Howe	45.7S	174.6E	39	Julia; American biographer, poet (1819-1910).
Hroswitha Patera	35.8N	34.8E	163	German writer (ca. 932-1002).
Hsueh T'ao	52.9S	13.7E	20	Chinese poet, artist (ca. A.D. 760).
Hua Mulan	86.8N	337.7E	23	Chinese warrior (ca. A.D. 590).
Huang Daopo	54.2S	165.1E	27	Chinese engineer.
Hull	59.4N	263.3E	48	Peggy; American war correspondent (1889-1967).
H'uraru Corona	9.0N	68.0E	150	Pawnee earth mother.
Hurston	77.7S	94.5E	65	Zora; American anthropologist, writer (ca. 1901-1960).
Hwangcini	6.3N	141.7E	30	Korean poet (16th century A.D.).
Hyndla Regio	22.5N	294.5E	2300	Norse wood giantess.
Ichikawa	61.6S	156.4E	36	Fusaye; Japanese feminist (1893-1981).
Idem-Kuva Corona	25.0N	358.0E	230	Finno-Ugraic harvest spirit.
Idunn Mons	46.5S	213.5E	250	Norse goddess.
Ilbis Fossae	71.9N	254.6E	512	Yakutian (Siberia) goddess of bloodshed.
Ilga	12.4S	307.4E	11	Latvian first name.
Ilithyia Mons	13.5S	315.5E	90	Greek goddess of childbirth.
Imdr Regio	43.0S	212.0E	1611	Norse giantess.
Inanna Corona	37.0S	35.9E	350	Semitic fertility goddess.
Inari Corona	18.0S	120.3E	300	Japanese rice goddess.
Indira	64.0N	289.5E	14	Hindu first name.
Indrani Corona	37.5S	70.5E	200	Hindu fertility goddess.
Ingrid	12.4S	308.8E	15	Scandanavian first name.
Inira	43.1S	239.2E	17	Eskimo first name.
Innini Mons	34.6S	328.5E	339	Babylonian earth mother worshipped at Kish.
Irene	49.8N	134.0E	13	First name from Greek.
Irina	34.8N	91.4E	22	Russian first name.
Iris Dorsa	52.7N	221.3E	2050	Greek goddess of rainbow.
Isabella	29.7S	204.1E	175	Of Castile; Spanish queen (1451-1504).
Isako	9.0S	277.9E	13	Japanese first name.
Ishtar Terra	70.4N	27.5E	5609	Babylonian goddess of love.
Isong Corona	12.0N	49.2E	540	Ibibio (Nigeria) fertility goddess.
Itzpapalotl Tessera	75.7N	317.6E	380	Aztec goddess of fate.
Ivka	68.0N	304.0E	16	Serbocroatian first name.
Iweridd Corona	21.0S	310.0E	500	Brythonic (English Celtic) earth goddess.
Ix Chel Chasma	10.0S	73.4E	503	Aztec wife of the sun god; probably moon goddess.
Iyele Dorsa	50.0N	278.7E	595	Moldavian witch who directed the winds.
Izumi Patera	50.3N	193.6E	74	Sikibu; Japanese writer (974-1036).
Jacqueline	70.0S	123.8E	17	First name from French.
Jadwiga	68.5N	91.0E	14	Polish first name.
Jael Mons	52.0N	121.0E	—	Hebrew goddess of dawn.

Feature name[a]	Lat. (°)	Long. (°)	Size (km)	Description
Javine Corona	5.5S	251.2E	450	Lithuanian harvest goddess.
Jeanne	39.9N	331.5E	27	French first name.
Jennifer	4.6S	99.8E	9	First name from Cornish.
Jerusha	22.0S	342.7E	17	Hebrew first name.
Jex-Blake	65.5N	169.0E	—	Sophia; British pioneer woman physician (1840-1912).
Jhirad	16.8S	105.6E	50	Jerusha; Indian physician.
Jocelyn	33.2S	276.4E	14	German first name.
Johanna	19.5N	247.2E	18	Hebrew first name.
Johnson	51.9N	254.5E	25	Amy; English aviator (1903-1941).
Joliot-Curie	1.6S	62.1E	80	Irene; French physicist, Nobel laureate (1897-1956).
Jord Corona	58.5S	349.5E	130	Norse earth goddess.
Joshee	5.5N	288.8E	34	Anandibai; Indian pioneer physician (1865-1887).
Juanita	62.9S	89.9E	19	Spanish first name.
Judith	29.1S	104.5E	20	Hebrew first name.
Julle	51.0N	242.5E	16	Czech, German first name.
Juno Chasma	30.5S	111.1E	915	Roman sky goddess; sister and consort of Jupiter.
Juno Dorsum	31.0S	95.6E	1652	As above.
Junkgowa Corona	37.0N	257.0E	400	Yulengor (Australia) fertility goddess.
Jurate Colles	56.8N	153.5E	418	Lithuanian sea goddess.
Kahlo	59.9S	178.8E	36	Frida; Mexican artist (1907-1954).
Kaikilani	32.7S	163.1E	19	First female ruler of Hawaii (ca. 1555).
Kaiwan Fluctus	48.0S	1.5E	1200	Ethiopian earth mother.
Kala	1.5N	314.2E	17	Kamchatka first name.
Kalaipahoa Linea	60.5S	338.0E	2400	Hawaiian war goddess.
Kallistos Vallis	51.1S	21.5E	900	Ancient Greek name for planet Venus.
Kamadhenu Corona	21.0N	136.5E	400	Hindu goddess of plenty.
Kamari Dorsa	59.2N	55.8E	589	Georgian sky maiden, daughter of weather god.
Kamui-Huci Corona	63.5S	322.5E	300	Ainu (Japan) earth goddess.
Kanik	32.6S	249.8E	16	Sakhalin first name.
Kara Linea	44.0S	306.0E	—	Icelandic valkryie.
Kartini	57.8N	333.0E	24	Raden Adjeng; Javanese educator (1879-1904).
Kauffman	49.5N	27.0E	24	Angelica; Swiss painter (1741-1807).
Kawelu Planitia	32.8N	246.5E	3910	Hawaiian mythological heroine, died and brought back to life.
Kayanu-Hime Corona	33.5N	57.0E	150	Shinto grain goddess.
Kaygus Chasmata	49.6N	52.1E	503	Ketian (Siberia) ruler of forest animals.
Kelea	8.9N	25.6E	25	Chieftess of Maui (ca. 1450).
Keller Patera	45.0N	273.5E	—	Helen; blind and deaf American lecturer (1880-1968).
Kelly	4.8S	359.2E	11	Gaelic first name.
Kemble	48.3N	14.4E	29	Frances (Fanny) Anne; English actress (1809-1893).
Kenny	44.3S	271.1E	55	Elizabeth; Australian nurse, therapist (1886-1952).
Khatun	40.3N	86.9E	37	Mihri; Turkish poet (1456-1514).

Feature name[a]	Lat. (°)	Long. (°)	Size (km)	Description
Khelifa	1.5S	129.8E	13	Arabic first name.
Khotun Corona	46.5S	81.5E	200	Yakut goddess of plenty.
Kingsley	22.6S	306.3E	24	Mary; English explorer, writer (1862-1900).
Kiris	20.9N	98.8E	13	Latvian first name.
Kitna	28.9S	277.3E	16	Kamchatka first name.
Klafsky	20.8S	188.0E	28	Katherina; Hungarian opera singer (1855-1896).
Klenova	78.1N	104.2E	125	Mariya; Soviet marine geologist (ca. 1910-1978).
Koidula	64.3N	139.1E	47	Lydia; Estonian poet (1843-1886).
Kollwitz	25.2N	133.6E	30	Käthe; German artist (1867-1945).
Konopnicka	14.5N	166.7E	20	Marie; Polish author (1842-1910).
Kottauer Patera	36.7N	39.6E	136	Helena; Austrian historical writer (1410-1471).
Kottravey Chasma	30.5N	76.8E	744	Dravidian (India) hunting goddess.
Kozhla-Ava Chasma	56.2N	50.6E	581	Marian (Volga Finn) forest goddess.
Krumine Corona	5.0S	261.5E	300	Lithuanian food goddess.
Kuanja Chasma	12.0S	99.5E	890	Mbundu goddess of the spirit of the hunt.
Kuan-Yin Corona	4.3S	10.0E	310	Chinese fertility goddess.
Kubebe Corona	15.5N	132.5E	125	Hittite mother earth goddess.
Kunapipi Corona	33.9S	86.0E	220	Australian mother earth goddess.
Kunhild Corona	19.3N	80.1E	200	German fertility maiden.
Kurukulla Mons	48.5N	103.0E	—	Etan goddess of wealth.
Kutue Tessera	39.5N	108.8E	653	Ulchian (Siberia) folklore toad that brings happiness.
La Fayette	70.2N	107.9E	68	Marie; French novelist (1634-1693).
Labé Patera	52.0N	273.0E	—	Etan (Tibet) goddess of wealth.
Lachappelle	26.7N	336.5E	35	Marie; French medical researcher (1769-1821).
Lachesis Tessera	44.4N	300.1E	664	Greek, one of three Fates.
Lada Terra	60.0S	20.0E	8614	Slavic goddess of love.
Lagerlöf	81.0N	285.3E	53	Selma; Swedish novelist (1858-1940).
Laima Tessera	55.0N	48.5E	971	Latvian and Lithuanian goddess of fate.
Lakshmi Planum	68.6N	339.3E	2343	Indian goddess of love and war.
Lampedo Linea	57.0N	295.0E	—	Scythian Amazon queen.
Landowska	84.5N	83.0E	45	Wanda; Polish pianist (1879-1959).
Langtry	17.0S	155.0E	52	Lillie; English actress (1853-1929).
Lasdona Chasma	69.3N	34.4E	697	Lithuanian main forest goddess.
Laulani	68.2S	121.3E	15	Hawaiian first name.
Laūma Dorsa	64.8N	190.4E	1517	Latvian witch who flies in the sky.
Laura	49.0N	141.0E	16	Spanish, Italian first name.
Laurencin	15.4S	46.4E	30	Marie; French painter (1885-1957).
Lavinia Planitia	47.3S	347.5E	2820	Roman; wife of Aeneas.
Lazarus	52.8S	127.2E	26	Emma; American poet (1849-1887).
Leah	34.2S	187.7E	15	Hebrew first name.
Lebedeva	45.2N	49.6E	42	Sarah; Russian sculptor (1881-1968).

Feature name[a]	Lat. (°)	Long. (°)	Size (km)	Description
Leda Planitia	44.0N	65.1E	2890	Mother of Helen, Castor and Pollux.
Ledoux Patera	9.2S	224.8E	75	Jeanne; French artist (1767-1840).
Lehmann	44.1S	38.7E	20	Inge; Danish geophysicist (1888-?).
Leida	23.3S	266.5E	22	Estonian first name.
Leila	44.3S	86.6E	19	First name from Arabic.
Lena	39.2N	22.8E	25	Russian first name.
Lenore	38.7N	292.3E	16	Greek first name (form of Helen).
Leonard	73.8S	185.0E	37	Wrexie; American assistant to P. Lowell (1867-1937).
Letitia	34.6N	288.6E	16	First name from Latin.
Leyster	1.0N	259.9E	45	Judith; Dutch painter (1609-1660).
Li Qingzhao	23.7N	94.3E	21	Chinese essayist, scholar (1085-1151).
Liban Farra	23.9S	353.5E	100	Irish water goddess.
Libera Corona	12.5N	24.0E	350	Roman fertility goddess.
Lida	29.1S	94.5E	14	First name from Greek.
Lida	36.5N	274.0E	—	Russian name.
Lilian	25.6N	336.0E	14	First name from Hebrew.
Lilinau Corona	34.0N	22.0E	200	Native American fertility maiden.
Liliya	30.0N	31.1E	18	Russian first name.
Lilwani Corona	29.5S	271.5E	500	Hittite earth goddess.
Lind	50.2N	354.9E	44	Jenny; Swedish singer (1820-1887).
Lineta	5.0S	354.1E	15	Latvian first name.
Lo Shen Valles	12.8S	89.6E	225	Chinese river goddess.
Lockwood	32.8S	51.5E	23	Belva; American lawyer, feminist (1830-1917).
Lois	17.9S	214.7E	15	First name from Greek.
Lonsdale	55.6N	222.1E	45	Kathleen; Irish physicist, crystallographer (1903-1971).
Loretta	19.7S	202.5E	13	First name from Latin.
Lotta	51.0N	336.0E	12	Swedish first name.
Louhi Planitia	80.5N	120.5E	2441	Karelo-Finn mother of the North.
Lucia	62.1S	67.8E	17	First name from Latin.
Lukelong Dorsa	73.3N	178.8E	1566	Polynesian goddess, creator of heavens.
Lullin	23.1N	81.0E	24	Maria; Swiss entomologist (1750-1831).
Lydia	10.7N	340.8E	15	First name from Greek.
Lyon	66.5S	270.5E	14	Mary; American educator, college president (1797-1849).
Lyudmila	62.0N	330.0E	16	Russian first name.
Ma Shouzhen	35.7S	92.4E	20	Chinese poet, painter (1592-1628).
Maa-Ling	14.7S	359.5E	6	Chinese first name.
Maan-Eno Corona	40.8N	102.5E	300	Estonian harvest goddess.
Maat Mons	0.5N	194.6E	395	Ancient Egyptian goddess of truth and justice.
MacDonald	30.0N	120.7E	19	Flora; Scottish heroine (1722-1790).
Madeleine	4.7S	293.2E	18	French first name.
Magda	67.0N	329.5E	12	Danish first name.

Feature name[a]	Lat. (°)	Long. (°)	Size (km)	Description
Magnani	58.0N	337.0E	30	Anna; Italian actress (1908-1973).
Mahina	2.0S	182.2E	16	Hawaiian first name.
Mahuea Tholus	37.5S	164.7E	110	Maori fire goddess.
Makh Corona	48.7S	85.0E	200	Assyro-Babylonian goddess of fecundity.
Makola	3.8S	106.7E	18	Hawaiian first name.
Malintzin Patera	57.0N	82.0E	—	(Malina); Aztec Indian guide, interpreter (1501-1550).
Maltby	23.3S	119.8E	40	Margaret; American physicist (1860-1944).
Mama-Allpa Corona	27.0S	31.0E	300	Peruvian harvest goddess.
Manto Fossae	64.5N	60.0E	—	Greek prophetess.
Manton	9.3N	26.8E	18	Sidnie; English zoologist (1902-1980).
Manzan-Gurme Tesserae	39.5N	179.5E	—	Ancestress who possesses the book of fate in mythology.
Manzolini	25.7N	91.1E	42	Anna; Italian anatomist, teacher (1716-1774).
Maram Corona	7.5S	221.5E	600	Oromo (Ethiopia) fertility goddess.
Maranda	4.9N	169.8E	14	Latvian first name.
Mardezh-Ava Dorsa	32.4N	68.6E	906	Marian (Volga Finn) wind goddess.
Margarita	12.8N	9.2E	13	Greek first name.
Margit	60.0N	273.0E	14	Hungarian first name.
Mari Corona	54.0N	151.0E	200	Cretan goddess of plenty.
Marie	21.7S	232.4E	15	French first name.
Maria Celeste	23.5N	140.5E	90	Daughter of Galileo (died 1634).
Markham	4.1S	155.6E	69	Beryl; English aviator (1902-1986).
Marsh	63.7S	46.7E	35	Ngaio; New Zealand playwright, novelist (1899-1982).
Martinez	11.7S	174.7E	25	Maria; Pueblo artist, potter (1886-1980).
Marzhan	58.9S	248.5E	20	Karakal first name.
Masako	30.2S	53.1E	26	Hozyo; Japanese ruler (1157-1225).
Maslenitsa Corona	77.0N	202.5E	—	Slavonic personification of fertility.
Mawu Corona	31.5N	241.0E	—	Fon (Benin) goddess of fertility.
Maxwell Montes	65.2N	3.3E	797	James C.; British physicist (1831-1879).
Maya Corona	23.0N	98.0E	225	Hindu mother earth goddess.
Mayaeul Corona	27.5S	154.0E	200	Mexican goddess of plenty.
Mbokomu Mons	15.1S	215.2E	460	Ngombe (Zaire) ancestor/goddess.
Mead	12.5N	57.4E	280	Margaret; American anthropologist (1901-1978).
Medeina Chasma	46.2N	89.3E	606	Lithuanian forest goddess.
Medhavi	19.5S	40.6E	30	Ramabai; East Indian author, humanitarian (1858-1922).
Megan	61.7S	130.6E	18	Welsh first name.
Meitner	55.9S	321.8E	98	Lise; Austrian physicist (1878-1968).
Melanie	62.8S	144.3E	16	First name from Greek.
Melba	4.7N	193.4E	2	Nellie; Australian opera singer (1861-1931).
Melia Mons	62.8N	119.3E	311	Greek nymph.
Mena Colles	52.5S	160.0E	850	Roman goddess of menses.
Menat Undae	24.8S	339.4E	100	Arabian desert goddess.

Feature name[a]	Lat. (°)	Long. (°)	Size (km)	Description
Meni Tessera	48.1N	77.9E	454	Semitic goddess of fate.
Mentha Mons	43.0N	237.4E	79	Roman goddess, personification of the human mind.
Merian	34.5N	76.2E	20	Maria; Dutch entomologist (1647-1717).
Merit Ptah	11.3N	115.7E	19	Egyptian queen, physician (ca. 2700 B.C.).
Mesca Corona	27.0N	342.6E	190	Irish fertility goddess.
Meskhent Tessera	65.8N	103.1E	1056	Egyptian goddess of fortune.
Metis Regio	72.0N	256.0E	729	Greek Titaness.
Metra Corona	26.0N	98.0E	—	Persian fertility/moon goddess.
Mežas-Mate Chasma	51.0N	50.7E	506	Latvian forest goddess.
Michelle	19.5S	40.4E	14	First name from French.
Milda Mons	52.5N	159.5E	—	Lithuanian goddess of love.
Millay	24.4N	110.9E	45	Edna St. Vincent; American poet (1892-1950).
Minerva Fossae	64.5N	252.5E	—	Roman goddess of war.
Mirabeau	1.1N	284.3E	22	Sibylle; French writer (died 1932).
Miralaidji Corona	14.0S	163.8E	300	Aborigine fertility goddess.
Miriam	36.5N	48.2E	15	First name from Hebrew.
Misne Chasma	77.1N	316.5E	610	Mansi (Siberia) forest maiden.
Mist Fossae	39.5N	247.3E	244	Norse Valkyrie.
Mnemosyne Regio	65.8N	277.9E	—	Greek Titaness.
Moira Tessera	58.7N	310.5E	361	Greek fate goddess.
Mokosha Mons	57.7N	255.0E	270	East Slavic main goddess.
Molpadia Linea	48.0S	359.0E		Amazon.
Molpe Colles	76.5N	195.0E	—	Greek; mother of Sirens.
Mona Lisa	25.6N	25.3E	80	Leonardo da Vinci's model, real name Lisa Giocondo (born ca. 1474).
Monika	72.5N	122.0E	24	German first name.
Montagu	36.9N	177.5E	20	Mary; English medical pioneer, poet, writer (1689-1762).
Montessori	59.1N	280.1E	43	Maria; Italian educator (1870-1952).
Montez	17.9N	266.6E	20	Lola; Irish dancer (1818-1861).
Moore	30.3S	248.3E	21	Marianne; American poet, editor (1887-1972).
Morana Chasma	68.9N	24.0E	317	Czech moon goddess.
Morisot	61.2S	211.4E	55	Berthe; French artist (1841-1895).
Morrigan Linea	54.5S	311.0E	3200	Celtic war goddess.
Moses	34.3N	120.1E	35	Anna "Grandma"; American painter (1860–1961).
Mots Chasma	51.9N	56.1E	464	Avarian (Caucasus) moon goddess.
Mowatt	14.7S	292.2E	40	Anna; American actress, playwright, author (1819-1870).
Mu Guiying	41.2N	80.7E	25	Chinese warrior.
Mukhina	29.5N	0.5E	24	Vera; Soviet sculptor (1889-1953).
Mumtaz-Mahal	30.3N	228.3E	39	Mogul empress for whom Taj Mahal was built (1592-1631).
Munter	15.3S	39.3E	36	Gabriele; German painter (1877-1962).
Muriel	41.7S	12.3E	19	First name from Celtic.
Muta Mons	56.0N	359.0E	—	Roman goddess of silence.

Feature name[a]	Lat. (°)	Long. (°)	Size (km)	Description
Mylitta Fluctus	56.0S	353.5E	1250	Semitic mother goddess.
Nabuzana Corona	8.5S	47.0E	69	Ganda (Uganda) crop goddess.
Nadine	7.8N	359.1E	19	First name from French.
Nadira	44.0N	201.7E	36	Uzbek poet (1791-1842).
Nagavonyi Corona	18.5S	259.0E	190	Ganda (Uganda) crop goddess.
Nalkowska	28.2N	290.0E	26	Zofia; Polish novelist, playwright (1884-1954).
Nambi Dorsum	72.5S	213.0E	1125	Ugandan sky goddess.
Nana	50.0N	75.0E	10	Serbocroatian first name.
Naomi	6.0N	70.1E	18	First name from Hebrew.
Natalia	67.0N	273.0E	10	Romanian first name.
Navka Planitia	8.1S	317.6E	2100	Arab mother-goddess Allat as goddess of good fortune.
Neago Fluctūs	48.9N	349.5E	—	Seneca (U.S.A.) goddess of silence.
Nefertiti Corona	35.8N	48.0E	—	Beautiful Egyptian queen (ca. 1390-ca. 1354 B.C.).
Nehalennia Corona	14.0N	10.0E	345	Teutonic fertility goddess.
Němcová	5.9N	125.0E	24	Božena; Czech novelist, poet (1820-1882).
Nemesis Tessera	45.9N	192.6E	355	Greek goddess of fate.
Nephele Dorsa	39.7N	139.0E	1937	Greek cloud goddess.
Nepret Corona	53.0N	7.0E	—	Egyptian grain goddess.
Nepthys Mons	33.0S	317.5E	350	Egyptian goddess of barren lands.
Nertus Tholus	61.2N	247.9E	66	German/Norse vegetation goddess.
Nevelson	35.3S	307.8E	75	Louise; Russian-born American artist (1899-1988).
Neyterkob Corona	49.5N	204.5E	—	Masai earth/fertility goddess.
Nightingale Corona	63.6N	129.5E	471	Florence; English nurse (1820-1910).
Nijinskaya	25.9N	122.3E	30	Bronislava; Russian dancer (1891-1972).
Nike Fossae	62.0S	347.0E	850	Greek goddess of victory.
Nilsson	76.0S	277.7E	26	Christine; Swedish opera singer, violinist (1843-1921).
Nin	3.9S	266.4E	27	Anaïs; French-born American novelist (1903-1977).
Nina	55.5S	238.7E	23	First name from Russian.
Ningal Undae	9.0N	60.7E	225	Sumerian desert goddess.
Ningyo Fluctus	5.5S	206.0E	970	Japanese fish goddess.
Ninhursag Corona	38.0S	23.5E	125	Babylonian earth goddess.
Nintu Corona	19.2N	123.5E	75	Akkadian earth goddess.
Niobe Planitia	21.0N	112.3E	5008	Greek; her 12 children were killed by Artemis and Apollo.
Nishtigri Corona	24.5S	72.0E	275	Hindu earth mother.
Nissaba Corona	25.5N	355.5E	300	Mesopotamian wisdom/fertility goddess.
Nofret	58.7S	252.0E	22	Egyptian queen (ca. 1900 B.C.).
Nokomis Montes	18.9N	189.9E	486	Algonquin (N. America) earth mother.
Noreen	33.5N	22.7E	20	Irish first name.
Noriko	5.3S	358.3E	7	Japanese first name.
Nsomeka Planitia	55.0S	170.0E	4500	Bantu culture heroine.
Nzingha Patera	69.0N	206.0E	—	(Ann Zingha) queen, head of Amazon band (1582-1663).

Feature name[a]	Lat. (°)	Long. (°)	Size (km)	Description
Oakley	29.3S	310.5E	22	Annie; American sharpshooter, entertainer (1860-1926).
Oanuava Corona	32.5S	255.5E	375	Gaulish Celtic earth goddess.
Obukhova	71.0N	289.0E	44	Nadezhda; Soviet singer (1886-1961).
O'Connor	26.0S	143.8E	27	Flannery; American novelist (1925-1964).
Odilia	81.5N	200.5E	20	Portuguese first name.
Oduduva Corona	11.0S	211.5E	150	Yoruba (Nigeria) fertility goddess.
Ohogetsu Corona	27.0S	85.7E	175	Japanese food goddess.
O'Keefe	24.5N	228.7E	76	Georgia; American artist (1887-1986).
Okipeta Dorsa	66.0N	238.5E	1200	Greek goddess of whirlwind.
Olesnicka	18.3N	210.8E	33	Zofia; Polish poet (flourished ca. 1550).
Olga	26.2N	283.8E	17	Russian first name.
Olwen Corona	37.5N	67.5E	175	Brythonic goddess of spring growth.
Olya	51.0N	292.0E	14	Russian first name.
Omeciuatl Corona	16.5N	119.0E	175	Aztec generative power.
Onatah Corona	49.0N	5.0E	—	Iroquois corn spirit.
Ops Corona	68.5N	89.0E	—	Greek fertility goddess.
Orczy	3.6N	52.2E	29	Emmuska; Hungarian novelist, playwright (1865-1947).
Orlova	56.5N	235.0E	28	Lyubov; Soviet actress (1902-1975).
Oshun Farra	4.2N	19.3E	80	Yoruba (Nigeria) fresh water goddess.
Osipenko	71.0N	321.0E	30	Polina; Soviet aviator (1907-1939).
Otau Corona	67.8N	298.7E	172	Bini (S. Nigeria) goddess of fertility.
Otygen Corona	57.0S	30.5E	400	Mongolian earth mother.
Ovda Fluctus	6.1S	95.5E	310	Named from regio where feature is located.
Ovda Regio	2.8S	85.6E	5280	Marijian (Russian) forest giantees.
Ozza Mons	4.5N	201.0E	507	Persian goddess honored by the Koreishies.
Pamela	11.1N	238.5E	13	English first name.
Pandrosos Dorsa	58.2N	206.2E	1254	Greek dew goddess.
Pani Corona	19.9N	231.5E	320	Maori fertility goddess.
Parga Chasma	24.5S	271.5E	—	Samoyed forest spirit.
Parra	20.5N	78.1E	50	Chilean writer.
Patti	34.8N	301.6E	40	Adelina; Italian singer (1843-1919).
Pavlova Corona	14.3N	38.9E	37	Anna; Russian ballerina (1885-1931).
Peck	29.0S	294.2E	30	Annie; American mountaineer, educator (1850-1935).
Peggy	20.4S	357.2E	12	English first name (form of Margaret).
Peña	23.6S	190.6E	32	Tonita (Quah Ah); Pueblo artist (1895-1949).
Penardun Linea	54.0S	344.0E	975	Celtic sky goddess.
Perchta Corona	17.0N	234.5E	500	German fertility goddess.
Phaedra	35.9N	252.7E	15	First name from Greek.
Phoebe Regio	6.0S	282.8E	2852	Greek Titaness.
Phra Naret Corona	66.6S	209.6E	150	Thai fertility goddess.
Phryne	46.2S	314.8E	40	Greek model, courtesan (fourth century B.C.).

Feature name[a]	Lat. (°)	Long. (°)	Size (km)	Description
Phyllis	12.3N	132.4E	13	First name from Greek.
Piaf	0.8N	5.1E	30	Edith; French singer, songwriter (1915-1963).
Piret	38.0N	42.0E	—	Estonian first name.
Piscopia	1.5N	190.9E	26	Elena; Italian mathematician, educator (1646-1684).
Pocahontas Patera	65.0N	49.5E	—	Daughter of Powhatan, Indian peacemaker (1595-1617).
Polina	42.2N	148.2E	24	Russian first name.
Pölöznitsa Corona	0.5N	302.0E	675	Finno-Ugric grain goddess.
Pomona Corona	79.3N	299.4E	315	Roman goddess of fruits.
Ponselle	63.0S	289.0E	53	Rosa; American opera singer (1897-1981).
Potanina	31.6N	53.1E	82	Aleksandra; Russian explorer (1843-1893).
Potter	7.2N	309.4E	52	Beatrix; English children's author (1866-1943).
Prichard	43.8N	11.1E	30	Katharine; Australian writer (1883-1969).
Purandhi Corona	26.1N	343.5E	170	Hindu goddess of plenty.
Qetesh Corona	20.5S	343.5E	80	Egyptian fertility goddess.
Quetzalpetlatl Corona	64.0S	354.5E	400	Aztec fertility goddess.
Quilla Chasma	23.7S	127.3E	973	Inca moon goddess.
Rachel	48.7S	13.5E	12	First name from Hebrew.
Radka	76.0N	95.0E	12	Bulgarian first name.
Raisa	27.5N	280.3E	13	Russian first name.
Rananeida Corona	62.6N	263.5E	448	Saami-Lapp goddess of spring and fertility.
Rand	63.8S	59.5E	27	Ayn; Russian-born American writer (1905-1982).
Rangrid Fossae	62.7N	356.4E	243	Norse Valkyrie.
Rani	64.5N	160.0E	12	Hindu first name.
Raskova Paterae	51.0S	222.8E	80	Marina M.; Russian aviator (1912-1943).
Rauni Corona	40.8N	271.9E	271	Finnish goddess of harvest, earth.
Razia Patera	46.2N	197.8E	157	Queen of Delhi Sultanate (India) (1236-1240).
Recamier	12.5S	57.9E	25	Jeanne-Françoise-Julie-Adélaïde; French patriot; (ca. 1777-ca. 1849).
Regina	29.8N	147.4E	35	First name from Latin.
Renenti Corona	32.7N	326.2E	200	Egyptian goddess of abundance.
Renpet Mons	76.0N	236.4E	138	Egyptian goddess of springtime and youth.
Rhea Mons	32.4N	282.2E	217	Greek Titaness.
Rhoda	11.5N	347.7E	13	First name from Greek.
Rhys	8.6N	298.8E	45	Jean; Welsh writer (1894-1979).
Richards	2.5N	196.0E	29	Ellen; founder of science of ecology (1842-1911).
Rigatona Corona	33.5S	278.5E	300	Celtic fertility goddess.
Riley	14.0N	72.2E	25	Margaretta; English botanist (1804-1899).
Rita	71.0N	335.0E	10	Italian first name.
Romanskaya	23.2N	178.4E	31	Sofia; Soviet astronomer (1886-1969).
Rosa Bonheur	9.8N	288.7E	105	French painter (1822-1899).
Rose	35.2S	248.2E	15	German first name.
Rosmerta Corona	0.0N	124.5E	300	Celtic fertility/luck goddess.

Feature name[a]	Lat. (°)	Long. (°)	Size (km)	Description
Rossetti	57.0N	7.0E	25	Christina; English poet (1830-1894).
Rowena	10.4N	171.3E	18	Celtic first name.
Roxanna	26.5N	334.6E	9	First name from Persian.
Rudneva	78.0N	176.0E	30	Varvara; Russian medical doctor (1844-1899).
Rusalka Planitia	9.8N	170.1E	3655	Russian mermaid.
Ruslanova	84.0N	16.0E	19	Lidiya; Soviet singer (1900-1973).
Ruth	43.2N	19.8E	18	Hebrew first name.
Sabin	38.5S	274.6E	36	Florence; American medical researcher (1871-1953).
Sabira	5.8S	239.9E	14	Tatar first name.
Sacajawea Patera	64.3N	335.4E	233	Blackfoot Indian woman who guided Lewis & Clark expedition to the Pacific Northwest (1786-1812).
Sachs Patera	49.0N	324.0E	—	Nelly; German-born Swedish playwright, poet (1891-1970).
Saga Vallis	76.1N	340.6E	450	Norse goddess in the form of a waterfall.
Salika	5.0S	97.7E	14	Mari first name.
Salme Dorsa	58.0N	28.0E	—	Estonian sky maiden.
Samantha	45.5N	281.4E	16	First name from Aramaic.
Samintang	39.0S	80.6E	24	16th century Korean poet.
Samundra Vallis	24.1S	347.1E	110	Indian river goddess.
Sand Patera	42.0N	15.5E	—	George (Aurore Dupin); French novelist (1804-1876).
Sandel	45.7S	211.6E	20	Cora; Norwegian author (1880-1974).
Sanger	33.8N	288.5E	84	Margaret; American medical researcher (1883-1966).
Sanija	33.1N	250.0E	18	Tatar first name.
Sapas Mons	8.5N	188.3E	217	Phoenician goddess.
Sappho Patera	14.1N	16.5E	92	Lyric poet; Lesbos, Asia Minor (flourished ca. 610-ca. 580 B.C.).
Sarah	42.4S	1.7E	19	Hebrew first name.
Sarpanitum Corona	52.3S	14.6E	170	Babylonian fertility goddess.
Sartika	63.4S	67.1E	28	Ibu Dewi; Indonesian educator (1884-1942).
Saskia	28.6S	337.2E	40	Artist's model, wife of Rembrandt.
Sati Vallis	3.2N	334.4E	225	Egyptian river goddess.
Saule Dorsa	58.0S	206.0E	1375	Lithuanian sun goddess.
Sayers	67.5S	230.0E	90	Dorothy L.; English novelist, playwright (1893-1957).
Scarpellini	23.4S	34.4E	25	Caterina; Italian astronomer (19th century).
Schumann-Heink Patera	74.0N	215.0E	—	Ernestine; German singer (1861-1936).
Sedna Planitia	42.7N	340.7E	3572	Eskimo; her fingers became seals and whales.
Seia Corona	3.0S	153.0E	225	Roman grain goddess.
Sekmet Mons	44.2N	240.8E	338	Ancient Egyptian goddess of war and battle.
Sel-Anya Dorsa	79.4N	81.3E	975	Hungarian wind goddess.
Selma	68.5N	156.0E	12	First name from Celtic.
Selu Corona	42.5S	6.0E	300	Cherokee corn goddess.
Semele Tholi	64.3N	202.9E	194	Phrygian (Phoenician) earth goddess.
Semuni Dorsa	75.9N	8.0E	514	Ulchian (Siberia) sky goddess.
Seoritsu Farra	30.0S	11.0E	230	Japanese stream goddess.

Feature name[a]	Lat. (°)	Long. (°)	Size (km)	Description
Seshat Mons	26.5N	33.0E	—	Egyptian goddess of writing.
Sévigné	52.5N	326.5E	30	Marie; French writer (1626-1696).
Seymour	18.2N	326.5E	65	Jane; English queen (ca. 1509-1537).
Shakira	3.0N	213.7E	19	Bashkir first name.
Shih Mai-Yu	18.4N	318.9E	25	Chinese physician (1873-1954).
Shimti Tessera	31.9N	97.7E	1275	Babylonian; Ishtar as the goddess of Fate.
Shiwanokia Corona	42.0S	279.8E	500	Zuni fertility goddess.
Sicasica Fluctus	52.0S	180.4E	175	Aymara (Bolivia) mountain goddess.
Siddons	61.6N	340.6E	47	Sarah; English actress (1755-1831).
Sidney	13.4N	199.6E	21	Mary; Elizabethan dramatist (1561-1621).
Sif Mons	22.0N	352.4E	200	Teutonic goddess, Thor's wife.
Sige Dorsa	32.0N	106.5E	—	Babylonian sky goddess.
Sigrid	63.5N	314.5E	20	Scandanavian first name.
Sigrun Fossae	52.3N	19.9E	971	Norse Valkyrie.
Simone	59.5N	82.0E	14	French first name.
Simonenko	26.9S	97.3E	35	Soviet astronomer.
Sinann Vallis	49.0S	270.0E	425	Irish river goddess.
Sirani	31.5S	230.4E	28	Elisabetta; Italian painter, etcher, printmaker (1638-1665).
Sith Corona	10.2S	176.5E	350	Norse harvest goddess.
Sitwell	16.7N	190.3E	35	Edith; English poet, critic (1887-1964).
Skadi Mons	64.0N	4.0E	40	Norse mountain goddess.
Snegurochka Planitia	86.6N	328.0E	2773	Snow maiden in Russian folktales, melted in spring.
Somagalags Corona	9.3N	348.5E	105	Bella Coola earth mother.
Sophia	28.7S	18.7E	17	First name from Greek.
Stanton	23.4S	199.9E	110	Elizabeth C.; American suffragist (1815-1902).
Stefania	51.0N	333.0E	12	Romanian first name.
Stein	30.0S	345.5E	24	Gertrude; American writer (1874-1946).
Steinbach	41.4S	256.8E	21	Sabina; German sculptor (ca. 1250).
Stina	37.2N	22.8E	38	Swedish first name.
Stopes Patera	42.5N	47.0E	—	Marie; English paleontologist (1880-1958).
Storni	9.8S	245.6E	25	Alfonsina; Argentine poet (1892-1938).
Stowe	43.2S	233.0E	82	Harriet B.; American novelist (1811-1896).
Stuart	30.8S	20.2E	67	Mary; Queen of Scots (1542-1587).
Suliko	9.6N	214.6E	19	Georgian first name.
Sullivan	1.3S	110.8E	22	Anne; American teacher of Helen Keller (1866-1936).
Sunrta Corona	8.3N	11.7E	170	Hindu fertility goddess.
Surija	5.3N	178.3E	15	Azerbaijani first name.
Susanna	6.0N	93.3E	16	First name from Hebrew.
Sveta	82.5N	271.0E	—	Russian first name.
Tacoma Corona	37.0S	288.0E	500	Earth goddess of Salish, Puyallup & Yakima Indians.
Taglioni	41.5N	122.8E	23	Maria; Italian ballet dancer (1804-1884).

Feature name[a]	Lat. (°)	Long. (°)	Size (km)	Description
Tai Shan Corona	32.5S	95.0E	175	Chinese fertility goddess.
Taira	1.5S	296.8E	19	Osset first name.
Takus Mana Corona	19.6S	345.3E	125	Hopi (USA) fertility goddess.
Talakin Mons	11.0S	355.4E	175	Navajo (USA) goddess.
Tamara	61.5N	317.5E	10	Georgian first name.
Tamfana Corona	36.3S	6.0E	400	Norse fertility goddess.
Tamiyo Corona	36.0S	297.5E	400	Japanese goddess of abundance.
Tanya	19.3S	282.7E	14	Russian first name.
Taranga Corona	16.5N	251.5E	525	Polynesian fertility goddess.
Tarbell Patera	58.2S	351.5E	80	Ida; American author, editor (1857-1944).
Tatyana	85.5N	217.0E	16	Russian first name.
Ta'urua Vallis	80.2S	247.5E	525	Tahitian word for the planet Venus.
Taussig	9.2S	228.9E	26	Helen; American pediatrician, heart researcher (1898-1986).
Teasdale Patera	67.6S	189.1E	75	Sara; American poet (1884-1933).
Tefnut Mons	38.6S	304.0E	182	Ancient Egyptian goddess of dew or rain.
Tellus Tessera	42.6N	76.8E	2329	Greek Titaness.
Tepev Mons	29.0N	44.3E	301	Quiche Mayan creator goddess.
Teresa	42.5S	9.9E	17	First name from Greek.
Teteoinnan Corona	38.5S	149.5E	125	Aztec fertility goddess.
Tethus Regio	66.0N	120.0E	—	Roman earth goddess.
Tey Patera	17.8S	349.1E	20	Josephine; Scottish author (1896-1952).
Tezan Dorsa	81.4N	47.1E	1079	Etruscan dawn goddess.
Thallo Mons	76.0N	233.5E	216	Greek goddess of flowering vegetation (Spring Hora).
Theia Mons	22.7N	281.0E	226	Greek Titaness.
Themis Regio	37.4S	284.2E	1811	Greek Titaness.
Thermuthis Corona	8.0S	33.0E	330	Egyptian fertility/harvest goddess.
Thetis Regio	11.4S	129.9E	2801	Greek Titaness.
Thomas	13.0S	272.6E	25	Martha; American college president (1857-1935).
Thouris Corona	6.5S	12.9E	190	Egyptian fertility goddess.
T'ien Hu Colles	30.8N	16.0E	—	Chinese sea goddess.
Tinatin Planitia	15.0S	15.0E	—	Georgian epic heroine.
Tipporah Patera	38.9N	43.0E	99	Hebrew medical scholar (1500 B.C.).
Tituba Patera	42.5N	214.0E	—	Nurse who started Salem witch hunt (ca. 1692).
Toklas	0.7N	273.2E	21	Alice; American writer, art patron (1877-1967).
Tomem Dorsa	31.2N	7.2E	970	Ketian (Siberia) Mother of the hot; lives in the sky, near the Sun.
Toyo-uke Corona	62.5S	41.5E	300	Shinto fertility goddess.
Trollope	54.8S	246.4E	26	Frances; English novelist (1780-1863).
Trotula Patera	41.3N	18.9E	146	Italian physician (A.D. 1097).
Truth	28.7N	287.7E	47	Sojourner; American abolitionist (1797-1883).
Tseraskaya	26.9N	78.8E	36	Lidiya; Soviet astronomer (1855-1931).
Tsiala	2.9N	100.0E	17	Georgian first name.

Feature name[a]	Lat. (°)	Long. (°)	Size (km)	Description
Tsvetayeva	64.0N	147.0E	40	Marina; Soviet poet (1892-1941).
Tubman	23.6N	204.5E	45	Harriet; American abolitionist (1820-1913).
Tumas Corona	16.3S	351.2E	200	Hopi (USA) fertility goddess.
Tünde	76.0N	197.0E	16	Hungarian first name.
Tusholi Corona	69.5N	101.2E	350	Chechen and Ingush (Caucasus) goddess of fertility.
Tussaud	21.8N	220.9E	24	Marie; Swiss wax artist (1760-1850).
Tuulikki Mons	10.3N	274.7E	520	Finnish wood goddess.
Udaltsova	20.3S	275.3E	28	Nadezhda; Russian artist (1885-1961).
Ukemochi Corona	39.0S	296.1E	300	Japanese fertility goddess.
Ulfrun Regio	20.5N	223.0E	3954	Norse giantess.
Ulrique	76.0N	55.5E	22	French first name.
Undset	52.0N	59.5E	28	Sigrid; Norwegian author (1882-1949).
Uni Dorsa	33.7N	114.3E	800	Etruscan goddess, same as Hera or Juno.
Uorsar Rupes	76.8N	341.2E	820	Adygan (Caucasus) goddess of hearth.
Upunusa Tholus	66.2N	252.4E	223	Earth goddess of Leti and Babar (Southwestern Islands, eastern Indonesia).
Ushas Mons	24.3S	324.6E	413	Indian goddess of dawn.
Ut Rupes	55.3N	321.9E	676	Siberian; Turco-Tatar goddess of the hearth fire.
Uvaysi	2.3N	198.2E	40	Uzbek poet (ca. 1780-ca. 1850).
Văcărescu	63.0S	199.6E	30	Helene; Rumanian poet, novelist (1866-1947).
Vacuna Corona	60.4N	96.0E	448	Sabinian (ancient Italy) goddess of harvest.
Vaiva Dorsum	53.2S	204.0E	520	Lithuanian rainbow goddess.
Vakarine Vallis	5.0N	336.4E	625	Lithuanian word for planet Venus.
Valadon	49.1S	167.5E	29	Suzanne; French painter (1869-1938).
Valborg	75.3N	272.0E	26	Danish first name.
Valentina	46.7N	143.2E	30	Latin first name.
Valkyrie Fossae	58.8N	7.5E	—	Norse battle maidens.
Vallija	26.4N	120.0E	16	Latvian first name.
Varma-Ava Dorsa	62.3N	267.7E	767	Mordvinian (Volga Finn) wind goddess.
Varz Chasma	71.3N	27.0E	346	Lezghin (Caucasus) moon goddess.
Vashti	6.8S	43.7E	18	Persian first name.
Vasudhara Corona	43.2N	2.7E	160	Buddhist female Bodhisattva of abundance.
Vedma Dorsa	49.8N	170.5E	3345	East Slav witch.
Vellamo Planitia	45.4N	149.1E	2154	Karelo-Finn mermaid.
Venilia Mons	32.7N	238.8E	320	Ancient Italian sea goddess.
Verdandi Corona	5.5S	65.2E	180	Norse bestower of blessings.
Veronica	38.1S	124.6E	16	First name from Latin.
Vesna	60.3S	220.4E	17	Slavic first name.
Vesta Rupes	58.3N	323.9E	788	Roman hearth goddess.
Vigée Lebrun	17.3N	141.3E	53	Marie; French painter (1755-1842).
Vihansa Linea	54.0N	20.0E	—	German war goddess.
Vinmara Planitia	53.8N	207.6E	1634	Swan maiden whom sea god Qat kept on Earth by hiding her wings.

Feature name[a]	Lat. (°)	Long. (°)	Size (km)	Description
Vir-ava Chasma	14.7S	124.1E	416	Mordvinian forest mother.
Vires-Akka Chasma	75.6N	341.6E	742	Saami-Lapp forest goddess.
Virginia	52.9S	185.9E	18	First name from Latin.
Virilis Tesserae	56.1N	239.7E	782	One of the names of Fortuna, Roman goddess of chance.
Virve	5.1S	346.8E	19	Estonian first name.
Volkova	75.1N	242.1E	52	Anna; Russian chemist (1800-1876).
von Paradis	32.2S	314.8E	36	Maria; Austrian pianist (1759-1834).
von Schuurman	5.0S	190.9E	29	Anna; Dutch linguist, writer, artist (1607-1678).
von Siebold	52.0S	36.7E	36	Regina; German physician, educator (1771-1849).
von Suttner	10.7S	234.9E	23	Bertha; Austrian journalist, pacifist (1843-1914).
Voynich	35.2N	56.0E	36	Lilian; English writer (1864-1960).
Wanda	71.5N	323.0E	16	Polish first name.
Wang Zhenyi	13.2N	217.8E	25	Chinese astronomer, geophysicist (18th century).
Warren	11.8S	176.5E	53	Mercy; American colonial poet, playwright, historian (1728-1814).
Weil	19.4N	283.1E	25	Simone; French author (1909-1943).
Wen Shu	5.0S	303.7E	33	Chinese painter (1595-1634).
West	26.1N	303.0E	29	Rebecca; Irish novelist, critic, actress (1892-1983).
Wharton	56.0N	62.0E	78	Edith; American writer (1862-1937).
Wheatley	16.6N	268.0E	75	Phillis; first black writer of note in America (1753-1785).
Whiting	6.0S	128.0E	36	Sarah; American physicist, astronomer (1847-1927).
Whitney	30.1S	151.3E	45	Mary; American astronomer (1847-1921).
Wieck	74.2S	211.0E	21	Clara; German pianist, composer (1819-1896).
Wilder	17.4N	122.4E	35	Laura Ingalls; American author (1867-1957).
Willard	24.7S	296.1E	47	Emma; American educator (1787-1870).
Winema	3.1N	168.6E	22	Modoc Indian heroine, peacemaker (ca. 1848-1932).
Winnemucca	15.4S	121.1E	30	Sarah; Piute interpreter, activist (ca. 1844-1891).
Wollstonecraft	39.2S	260.7E	44	Mary; English author (1759-1797).
Woodhull Patera	37.5N	306.0E	—	Victoria; American-English lecturer (1838-1927).
Woolf	37.7S	27.1E	25	Virginia; British writer (1882-1941).
Workman	12.9S	299.9E	19	Fanny; American mountaineer, author (1859-1925).
Wu Hou	25.4S	317.4E	30	Chinese empress (ca. 624-705).
Wurunsemu Tholus	40.6N	209.9E	83	One of the main figures in Hatti (proto-Hittite) mythology, sun goddess and mother of gods.
Xanthippe	10.8S	11.7E	41	Wife of Socrates (5th century B.C.).
Xiao Hong	43.6S	101.5E	37	Chinese novelist (1911-1942).
Xochiquetzal Mons	3.5N	270.0E	80	Aztec goddess of flowers.
Yablochkina	48.6N	195.5E	63	Aleksandra; Soviet actress (1866-1964).
Yale	13.4S	271.2E	20	Caroline; American educator of the deaf (1848-1933).
Yaroslavna Patera	38.8N	21.2E	112	Russian, wife of Prince Igor (12th century).
Ymoja Vallis	71.6S	204.8E	390	Yoruba (Nigeria) river goddess.
Yonge	14.0S	115.1E	26	Charlotte; English writer (1823-1901).

Feature name[a]	Lat. (°)	Long. (°)	Size (km)	Description
Yoshioka	32.4S	58.8E	20	Yayoi; Japanese physician, college founder (ca. 1871-1959).
Yumyn-Udyr Dorsa	78.0N	130.0E	—	Marian (Volga Finn) daughter of main god.
Yvonne	56.0S	298.3E	15	French first name.
Zamudio	9.6N	189.2E	19	Adela; Bolivian poet (1854-1928).
Zdravka	65.0N	299.0E	12	Bulgarian first name.
Zemina Corona	11.7S	186.0E	530	Lithuanian fertility goddess.
Zenobia	29.3S	28.5E	39	Queen of Palmyra (Syria) (third century A.D.).
Zhilova	66.3N	125.4E	45	Maria; Russian astronomer (1870-1934).
Zhu Shuzhen	26.5S	356.6E	32	Chinese poet (1126-1200).
Zija	3.5S	265.0E	18	Azerbaijani first name.
Zina	41.9N	319.9E	15	Romanian first name.
Zisa Corona	12.0N	221.0E	850	German harvest goddess.
Zlata	64.5N	334.0E	8	Serbocroatian first name.
Zorile Dorsa	39.9N	338.4E	1041	Moldavian dawn goddess.
Zorya Tholus	9.4S	335.3E	22	Slavic dawn goddess.
Zoya	68.0N	237.0E	22	Russian first name.
Zvereva	45.2N	282.9E	44	Lidiya; Russian aviator (1890-1916).

APPENDIX 6 **REFERENCE MAP OF VENUS**

The map shows the geographical locations of various features mentioned in the text. Lowlands are indicated by oblique ruling, volcanic plains are white areas, and highland massifs are shown in shades of orange with the highest mountains indicated in solid orange.

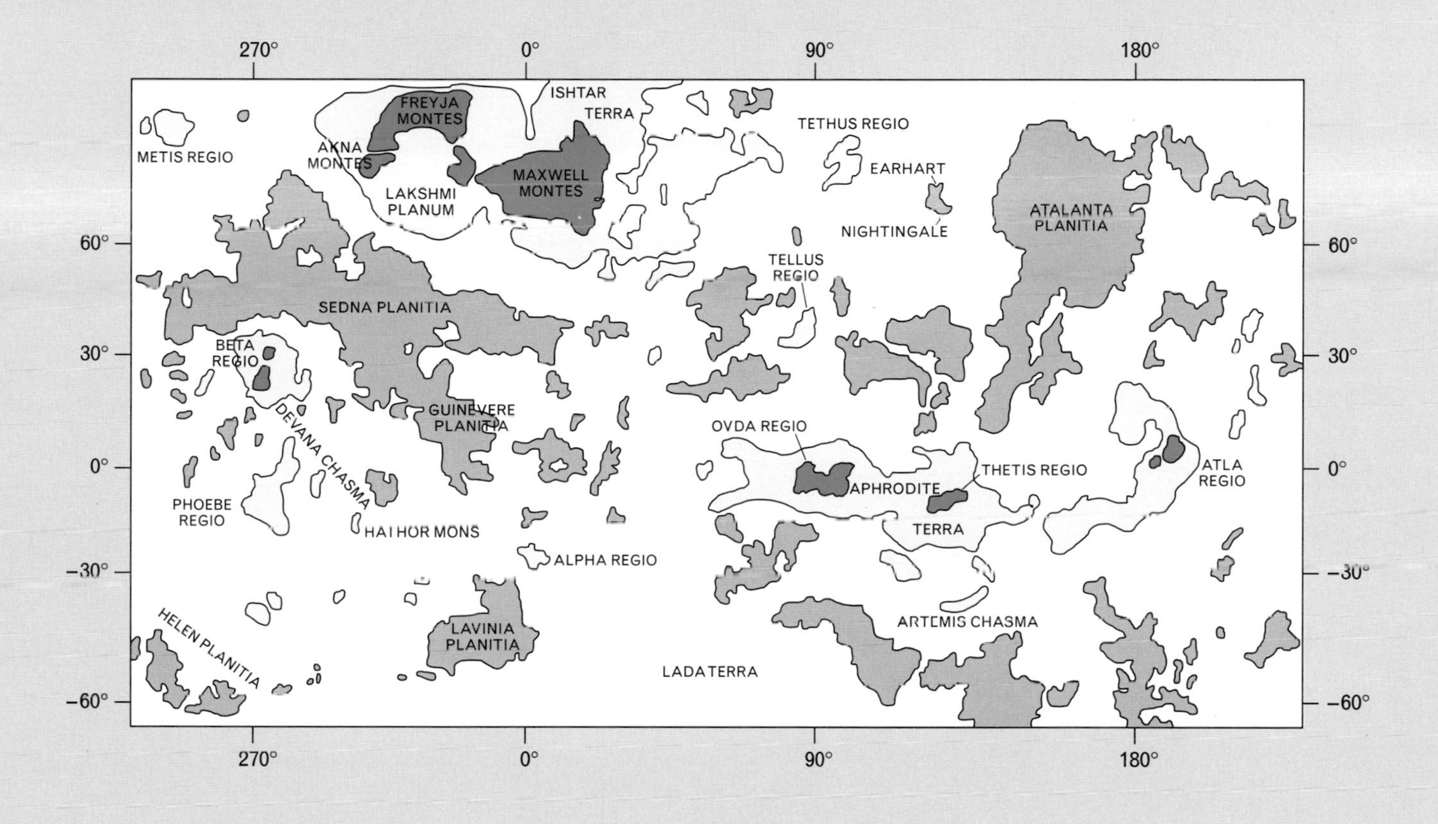

SELECTED BIBLIOGRAPHY

Bianchini, F. *Observations concerning the Planet Venus (1725).* Translated by Sally Beaumont. Springer-Verlag, Berlin/Heidelberg/New York1995.

Cattermole, P. J. *Venus – the Geological Story.* University College London Press, 1994.

Cattermole, P. J. *Planetary Volcanism* (2nd edition). Praxis Publishing/Wiley, Chichester, 1995.

Frankel, C. *Volcanoes of the Solar System.* Cambridge University Press, Cambridge, 1996.

Greeley, R. *Planetary Landscapes* (2nd edition). Chapman and Hall, London 1994.

Greeley, R. and Batson, R. (eds.) *The NASA Atlas of the Solar System.* Cambridge University Press, Cambridge, 1997.

Head, J. W., Crumpler, L. S. and Aubele, C. J. Venus volcanism: classification of volcanic features and structures, associations and global distribution of volcanic features and structures, associations and global distribution from Magellan data. *J. Geophys. Res.*, **97**, 13153–97, 1992.

Hunt, G. and Moore, P. *The Planet Venus.* Faber and Faber, London 1982.

Hunten, D. M., Colin, L., Donahue, D. M. and Moroz, V. I. (editors). *Venus.* University of Arizona Press, Tucson, 1983.

Moore, Patrick. *Mission to the Planets.* Cassell, London 1995.

NASA. *Venus – Magellan – the Unveiling of Venus.* JPL publication 400-345 3/89, Pasadena.

NASA. *Magellan – Revealing the Face of Venus.* JPL Publication 400-484 3/93, Pasadena.

Schaber, G. G. *et al.* Geology and distribution of impact craters on Venus: What are they telling us? *J. Geophys. Res.*, **97**, 13257–302, 1992.

Stofan, E. R. The new face of Venus. *Sky and Telescope*, August 1993. (Includes 3D views of Venus.)

Stofan, E. R. *et al.* Global distributions and characteristics of coronae and related features on Venus: implications for origin and relation to mantle processes. *J. Geophys. Res.*, **97**, 13347–78, 1992.

Woolf, H. *The Transits of Venus.* Princeton University Press, 1959.

INDEX

General items

Formations on Venus referred to in the text